Stochastic Lagrangian Modeling for Large Eddy Simulation of Dispersed Turbulent Two-Phase Flows

Authored By

Abdallah Sofiane Berrouk

Chemical Engineering Department

Petroleum Institute, Abu Dhabi

United Arab Emirates

DEDICATION

To my parents, sisters Houda & Mouna, brother Tarek and wife Karima

CONTENTS

FOREWORD

ENDORSEMENT

Dispersed flows with droplets and particles abound in nature from clouds, mist and fogs to the long-range transport of fine dust released in desert storms or in volcanic eruptions. They control the weather and influence the climate. They play key roles in many industrial energy processes – from spray drying, pneumatic conveying and fluidized beds, to coal gasification and mixing and combustion processes. They can have a profound effect on our health and quality of life, e.g. the inhalation of very fine air-borne particulate (PM10s) damages respiratory functions, and lead to increased cardio- pulmonary mortality rates and allergic disorders. Understanding the behavior of these flows, through modeling and experiment, is therefore important in our control of the environment, improving our health, and in the design and improvement of industrial processes.

There are essentially two ways of modeling dispersed flows, the Euler-Lagrange (E-L) approach where individual particle are tracked through an Eulerian flow field or an entirely Eulerian (E-E) approach where both the dispersed and continuous phases are described by a set of continuum equations that represent the conservation of mass moment and energy within an elemental volume of the mixture. This e-Book is about the E-L approach and exploits the recent advances in modeling complex flows in which LES has been used to describe the underlying large scale features of the continuous phase and stochastic equations based on the Generalized Langevin equations for the sub-filtered (SGS) motion of the fluid seen by the particle. The main advantage of this approach is that one can deal very successfully with particle dispersion and transport in complex industrial flows with complex geometries and boundary conditions, noting also that much success has been achieved using this approach in more generic flows like deposition of particles in a turbulent boundary. The author has been successful in presenting a thoroughly comprehensible and readable text that engineering researchers can use as a source of reference as well as for fundamental understanding. In order to achieve the dual objectives of understanding and simplicity, lengthy mathematical analysis has been avoided and emphasis has been placed on detailed applications. In view of the recent advances in LES and sub grid scale modeling and the application of stochastic methods, a e-Book of this nature is very timely and one that I would thoroughly recommend to all young researchers embarking on a study of industrial dispersed flows.

Michael W. Reeks

Professor of Multiphase Flows
Newcastle University
U.K

PREFACE

This e-Book presents an introduction to the numerical simulation of turbulent dilute two-phase flows using the very promising mesoscale approach namely; Large Eddy simulation (LES). The text details the use of the stochastic diffusion process in conjunction with LES approach to simulate industrial applications dealing with very small, turbulence-responsive particles convected by high-Reynolds, non-homogeneous and anisotropic flows. This novel approach has been proven to increase the accuracy of two-phase LES while keeping the computing cost at an affordable level.

The goal is to present a text that engineering researchers can read and understand and from which they can attack a variety of industrial problems. It serves only as a preliminary exposure to the readers who are interested in developing the subject at more advanced level. In order to achieve the dual objectives of understanding and simplicity, lengthy mathematical analysis was avoided and emphasis was placed on detailed applications.

The material included in this book is organized to facilitate the understanding of the numerical approach in the context of dilute gas-particle systems. The context in which the stochastic LES-particle is developed is explained in Chapter **1**. In Chapters **2** and **3**, the most important aspects of LES of single- and two-phase turbulent flows are covered, respectively. A separate chapter (Chapter **4**) details the stochastic model developed to track small inertial particles transported by the LES-predicted turbulent flows. For that purpose, the simplest mathematical approach is used, consistent with technical vigor. In the subsequent three chapters (Chapters **5**, **6** and **7**), three important flow applications are presented in almost every details. The aim is to show the promising potential which the stochastic LES-particle approach has in tackling dispersed industrial flows. Also, to use the stochastic LES-particle approach to explain many of the experimental findings on inertial particle transport by turbulent flows.

The e-Book is written with the necessary dose of both the physics and the numerics needed to quickly understand such a topic, and use it to tackle important industrial applications. It is brief and easy, yet to-the-point, description of an important topic written in a style and language often preferred by young graduate students and young researchers. It may be used in several ways at various stages of particulate flow modeling. It may help postgraduates and researchers interested in applying tractable yet powerful numerical tools to solve problems involving multiphase flows. Researchers in chemical, mechanical, petroleum and environmental engineering are the primary targets. Also, R&D people working for industries, such as the chemical and petrochemical industries should find this book very helpful in their continuing endeavor to adopt shorter design cycles through increased reliance on numerical prediction. Conventional models for turbulent dispersed flows do not appear capable of meeting the growing needs of these industries in this regard and the physical testing is prohibitive.

This e-Book is essentially the outcome of my last five years of association with this subject. I have received great deal of help from numerous persons over these years in formulating and revising my views on this numerical approach for particulate turbulent flows. I am particularly indebted to my teacher and mentor, Professor D. Laurence, who has been one of the leading practitioners of large eddy simulation technique for three decades at both Electricite de France and the University of Manchester. There are no adequate words to express his contributions to my understanding of the subject. I was fortunate to have an opportunity to work with Professor J.J. Riley at University of Washington-Seattle. Without his support and critics, it would not have been possible to develop the stochastic model. I would like to acknowledge the support provided by Emeritus Professor M.W. Reeks of University of Newcastle, Emeritus Professor D.E. Stock of Washington State University, and Professor Krishnanswamy Nandakumar of Louisiana State University. I would like to thank Dr. Alexander Douce of Electricite de France, Dr. Andy Lowe of the University of Manchester, and Dr. Chunliang Wu of Guangdong Ocean University, China for sharing knowledge about the subject.

I am grateful to one of my brilliant students, Mohamed Abu-Taqiya, at the Petroleum Institute- Abu Dhabi who helped me collect the required information, edit and proofread the book. Any remaining errors or shortcomings are the responsibility of the author. Finally, I wish to thank my wife, Karima, for her patience, understanding and enthusiastic support, which carried me through this long and arduous writing process.

Abdallah Sofiane Berrouk

Chemical Engineering Department
Petroleum Institute
UAE
Abu Dhabi, January 2011

CHAPTER 1

Introduction

Abstract: Understanding the dispersion and the deposition of inertial particles in turbulent flows is a domain of research of utmost practical interest. With advances in computing resources, Direct Numerical Simulation (DNS) and Large Eddy Simulation (LES) have become powerful tools for the investigation of particle-laden turbulent flows with the hybrid Eulerian-Lagrangian approach playing a key role in predicting inertial particle dispersion and deposition. Computational intractability that arises due to the need of solving all the scales has restricted DNS to the very low Reynolds number turbulent flows that are not often of practical interest. LES, by solving only the large energy-containing eddies and modeling the small quasi-universal scales, is relaxed from this restriction. Thus, tackling high Reynolds number turbulent flow becomes possible. The use of large-eddy simulation has increased over the years as a promising tool to address these types of problems with the required accuracy at an affordable computing cost. In LES of dispersed turbulent multiphase flows, it has been common that tracking inertial particles in turbulent flows is carried out using only the filtered velocity field. This turned out to be inaccurate for cases dealing with very small, turbulence-responsive particles. For these cases, the time-dependent velocity field seen by the inertial particles can be stochastically constructed in a Lagrangian framework. This can be achieved through the use of a stochastic diffusion process such as Langevin models.

Keywords: Large eddy simulation; turbulent flows; Eulerian-Lagrangian; sub-grid scales; filtered velocity; stochastic process; diffusion; dispersion; deposition; inertial particles.

Understanding the dispersion and the deposition of inertial particles in turbulent flows is a domain of research of utmost practical interest. Inertial particle transport and dispersion are encountered in a wide range of flow configurations, whether they are of industrial or environmental character. Indeed, predicting the distribution of particles after injection into combustion chambers or those travelling in gas streams in pipes plays a key role in the efficiency and stability of energy conversion and transport processes. Also, the effectiveness of natural flow phenomena for dispersing particulate pollutants and aerosols to acceptable concentration levels is another issue of interest for environmental and military purposes. In most of these applications, focus is put on how particles are dispersed by the turbulent motion and where the particles end up. When the particle mass-loading is high, it is important also to see how the presence of particles can modulate the turbulence.

With advances in computing resources, Direct Numerical Simulation (DNS) and Large Eddy Simulation (LES) have become powerful tools for the investigation of particle-laden turbulent flows with the hybrid Eulerian-Lagrangian approach playing a key role in predicting inertial particle dispersion and deposition. Computational intractability that arises due to the need of solving all the scales has restricted DNS to the very low Reynolds number turbulent flows that are not often of practical interest. LES, by solving only the large energy-containing eddies and modeling the small quasi-universal scales, is relaxed from this restriction. Thus, tackling high Reynolds number turbulent flow becomes possible. For many realistic applications with complex physics involving one or more phases, the Reynolds-Averaged Navier-Stokes approach (RANS) has been proved inherently ill-posed and it is facing important limitations [1]. Thus, the use of large-eddy simulation has increased over the years as a promising tool to address these types of problems with the required accuracy at an affordable computing cost.

Scales separation in large eddy simulation is carried out through filtering. A subfilter or subgrid scale model (SGS) is designed to account for the interaction between the LES-resolved and the unresolved scales. Thus the instantaneous information concerning the dynamics of the small scales is washed out. In LES of dispersed turbulent multiphase flows, it has been common that tracking inertial particles in turbulent flows is carried out using only the filtered velocity field. Therefore, any transport by the subfilter fluctuations is considered negligible. In this case, the fluid velocity seen by inertial particles along their trajectories is assumed to be the Eulerian filtered velocity. This should be a justifiable assumption in most light loading applications in which large Stokes-number inertial particles are tracked numerically on very fine LES grids using small filters. In these cases, the inertial particles do not sense turbulent fluctuations associated with the small or subfilter scales. Also, small amount of the turbulent kinetic energy of the larger scales is filtered out. Thus, the two-phase LES predictions accuracy would solely hinge on the aptness of the subfilter model to include the effects of the unresolved scales on the large scales when computing the single-phase velocity field.

Abdallah Sofiane Berrouk

Unfortunately, this may not be enough for cases dealing with very small, turbulence-responsive particles, for example, in the case of aerosols in high Reynolds number wall-bounded and/or highly non-equilibrium turbulent flows. Often in such cases, the LES grid is not fine enough, in particular, near the walls and in zones of recirculation and boundary layer detachment. All of this is dictated by the computing resource restriction. Thus, a significant amount of turbulent kinetic energy may be filtered out and the discarded subfilter fluctuations can be the major source of errors in LES predictions of both carrier and dispersed phase statistics. Also, neglect of subfilter scale motion can lead to inaccurate turbulence modulation predictions in situations where the particle mass loading is high. Indeed, the knowledge of the subfilter motion is important when the common applied point-force model is used to account for the particle effects on turbulence [2].

For these cases, the instantaneous velocity can be synthetically derived from the LES velocity field. This can be accomplished in many ways. One way is to add an unsteady subfilter component directly to the filtered velocity provided by LES. The subfilter fluctuations can be obtained by solving a transport equation for the subfilter or residual kinetic energy [3] and using a random number generated using a Gaussian distribution. Another approach consists of approximately de-filtering the LES velocities to generate the fluid phase instantaneous field to use to track inertial particles [4, 5]. It is also possible to add a stochastic forcing to the filtered Navier-Stoke equation [6]. The amplitude of this forcing is based on the subfilter turbulence scales as predicted by the SGS model used to account for the effects of the small scales on the large scales. For these approaches, the DNS-like velocity field is reproduced in an Eulerian framework and used to track inertial particles in a Lagrangian reference. Another way is to stochastically construct the time-dependent velocity field seen by the inertial particles in a Lagrangian framework. This can be achieved through the use of a stochastic diffusion process such as Langevin models [7]. This Lagrangian approach has been extensively used in the framework of RANS to construct total turbulence fluctuations based on the mean flow statistics [7, 8].

There is a body of research on LES of dispersed turbulent flows in simple configurations that has appeared in numerous publications over the last two decades [9, 10, and 11]. The purpose was to study the physics of these particulate systems. More recently these investigations have been extended to include simulations of dispersed flows in applications of practical interest following the success of single-phase LES [12, 13]. Few of these large eddy simulations have included the effects of the subfilter fluctuations on the inertial particle transport [14-18]. With the emergence of massively parallel computing and robust unstructured mesh codes, performing LES in realistic configurations such as combustors has become possible [12, 13].

Considerable experimental work on turbulent particle-laden flows has also been performed. However, in most of the experiments, only large diameter particles were considered. We mention herein the experimental works of Arnason and Stock [19] Sommerfeld and Qui [20,21] and Pui *et al.* [22] who considered very small particles in confined turbulent flows.

The early works on stochastic LES-particle approach made use of Langevin equation to account for the transport by subfilter fluctuations [23]. Promising results were obtained for particles with very small response times in a decaying isotropic turbulent flow. However, two physical phenomena linked to inertial particles dispersion, namely particle inertia and cross trajectory effects were neglected. This is valid only for inertial particles with vanishing response times. Improvements to the approach have followed to account for the particle inertial character and body forces in the context of the much complicated non-homogeneous and anisotropic turbulence. The stochastic modeling also proved successful in predicting particle Lagrangian velocity increments [24]. The model is capable of handling the far from Gaussian behavior of particle acceleration statistics in a manner consistent with the experimental findings.

This book explains the use of the stochastic Lagrangian model based on Langevin equation to model the fluid velocity seen by inertial particles for large eddy simulation of high Reynolds-number, non-homogeneous and anisotropic flows. By achieving that, the model accounts for the subfilter motion on inertial particle transport since the LES filtered velocity is known. Thus, modeling the fluid velocity as seen by inertial particles or modeling the subfilter motion is interchangeable. Corrections for inertia and cross trajectory effects are included to account for the dynamic response of inertial particles to turbulent flow different from fluid elements. It is aimed that this model will be used as an integral part of Eulerian-Lagrangian large eddy simulation of particle-laden turbulent flows of practical interest. This approach is motivated by the current trend in the industry toward a shorter design cycles through increased reliance on numerical prediction. RANS models for turbulent dispersed flows do not always appear capable of meeting the needs of industry in this regard. Physical testing is prohibitive. Consequently, a complete large eddy simulation appears to be the best prediction tool that can meet those expectations.

REFERENCES

[1]   Lakehal D. On the modeling of multiphase turbulent flows for environmental and hydrodynamic applications. Int J Multiphase Flow 2002; 28; 823-863.

[2]   Durbin PA, Pettersson Reif BA. Statistical theory and modeling for turbulent flows. Chichester: John Wiley & Sons 2003.

[3]   Sarghini F, Piomelli U, Balaras E. Scale-similar models for large-eddy simulations. Phys Fluids 1999; A(11)6.

[4]   Kleinstreuer C. Two-phase flow: theory and applications. Taylor and Francis Group Inc 2003.

[5]   Scotti A, Meneveau C. Fractal model for course-grained nonlinear partial differential equations Phys Rev Lett 1997; 78; 867.

[6]   Durbin PA, Pettersson Reif BA. Statistical theory and modeling for turbulent flows. Chichester: John Wiley & Sons 2003.

[7]   Maxey MR, Riley JJ. Equation of motion for a small rigid sphere in a nonuniform flow. Phys Fluids 1983; 26(4); 883-9.

[8]   Piomelli U, Ferziger JH, Moin P. New approximate boundary conditions for large eddy simulations of wall-bounded flows. Phys Fluids 1989 A(1)6.

[9]   Piomelli U. High Reynolds number calculations using the dynamic subgrid scale stress model. Phys. Fluids 1993; A5 (6).

[10]   Rizk MA, Elghobashi SE. Two-equation turbulence model for dispersed dilute confined two-phase flows. Int J Multiphase Flow 1989; 15; 119-33.

[11]   Simonin O, Deutsch E, Boivin M. Large eddy simulation and second-moment closure model of particle fluctuating motion in two-phase turbulent shear flows. In turbulent shear flows 9. eds. F. Durst, N. Kasaki, B. Launder, F. Schmidt, and J. Whitelaw Springer Verlag 1995; 85-115 pp.

[12]   Apte SV, Mahesh K, Moin P, Oefelein JC. Large-eddy simulation of swirling particle-laden flows in a coaxial-jet combustor. Int J Multiphase Flow 2003; 29; 1311-1331.

[13]   Sankaran V, Menon S. LES of spray combustion in swirling flows. J Turbulence 2002; 3(11)

[14]   Amiri AE, Hannani SK, Mashayek F. Large eddy simulation of heavy particle transport in turbulent channel flow. Numerical Heat Trans 2006; 50(Part B); 285-313.

[15]   Armenio V, Piomelli U, Fiorotto V. Effect of the subgrid scales on particle motion. Phys Fluids 1999; A(11); 3030-3042.

[16]   Okongo'o N, Bellan J. Consistent Large-eddy simulation of a temporal mixing layer laden with evaporating drops. Part 1 Direct numerical simulation, formulation and a priori analysis. J Fluids Mech 2004; 499; 1-47.

[17]   Segura JC. Ph.D. thesis. ME Dept. Stanford University. California 2004.

[18]   Wang Q, Squires KD. Large eddy simulation of particle-laden turbulent channel flow. Phys Fluids 1996; 8(5); 667-683.

[19]   Arnason G, Stock DE. Dispersion of particles in turbulent pipe flow. American Society of Mechanical Engineers. Fluid Engin Div 1983; 10; 25-29.

[20]   Sommerfeld M, Qiu HH. Detailed measurements in a swirling particulate two-phase flow by a phase-Doppler anemometer. Int J Heat Fluid Flow 1991; 12; 20-28.

[21]   Sommerfeld M, Qiu HH. Characterization of particle-laden, confined swirling flows by a phase-doppler anemometer. Int J Heat Fluid Flow 1993; 19; 1093-1127.

[22]   Pui DYH, Romay-Novas F, and Liu BYH. Experimental study of particle deposition in bend of circular cross section. Aerosol Sci Technol 1987; 7; 301-305.

[23]   Shotorban B, Mashayek F. A stochastic model for particle motion in large-eddy simulation. J Turbulence 2006; 7; 18,1-13.

[24]   Bini M, Jones WP. Particle acceleration in turbulent flows: A class of nonlinear stochastic models for intermittency. Phys Fluids 2007; 19(035104).

CHAPTER 2

Large Eddy Simulation of Single-Phase Flow

Abstarct For large eddy simulation of particle-laden turbulent flows, the accuracy with which the continuous phase is predicted is crucial for the simulation of the dispersed phase. In this chapter, issues that are important for LES of single-phase turbulent flow, including filtering, SGS modeling and numerics, are reviewed. Also, the advantage that LES holds over DNS and RANS to simulate practical single-phase turbulent flows is discussed.

The LES philosophy is based on the trivial observation that the loss of information incurred during filtering places a limit on the achievable accuracy of LES. Keeping that in mind, a broad array of SGS models for LES purposes have been developed using a variety of different approximations. Such models have been successful in many flows, yet they all have shortcomings.

There are many important issues in LES which make the validation of the results, and of the SGS models themselves, essential. The most obvious ones are those related to the physical modeling of the subgrid stresses, which is in many cases based on engineering approximations, justified by little more than dimensional considerations. Numerical issues are also important, since most models use as input the velocity gradients of the filtered field, which depend on the smallest scales resolved by the simulations, and therefore strongly influenced by numerical errors. Consistency with filtering of the physical models and numerical schemes adopted is as important as the previous issues.

Keywords: Large eddy simulation; turbulent flows; single-phase flow; filtering; sub-grid scales; eddy-viscosity models; scale-similar models; momentum conservation; consistency; numerical errors.

2.1. INTRODUCTION

For large eddy simulation of particle-laden turbulent flows, the accuracy with which the continuous phase is predicted is crucial for the simulation of the dispersed phase. In this chapter, issues that are important for LES of single-phase turbulent flow, including filtering, SGS modeling and numerics, are reviewed. Also, the advantage that LES holds over DNS and RANS to simulate practical single-phase turbulent flows is discussed. We start with the definition of LES strategy.

Large Eddy Simulation (LES) is essentially an under-resolved simulation of the complex turbulence phenomenon. It uses a model to account for the lack of small-scale resolution. In LES the conflicting requirements of complexity reduction while maintaining accurate predictions are achieved by coarsening the numerical description on one hand and using a subgrid stress (SGS) model on the other hand. Spatial filtering, which externally specifies the physical details that will ideally be retained in the LES solution, is used to approach the coarsening issue. Maintaining the dynamical properties of the resolved large scales is tackled by introducing subgrid scale (SGS) modeling. This modeling deals with the closure problem that arises from filtering the nonlinear term in the system of the partial differential equations that govern the turbulent flows.

With these objectives, LES has emerged as an intermediate approximation between two approaches: Direct Numerical Simulation (DNS) and Reynolds-Averaged Navier Stokes (RANS). For DNS, all the scales of the turbulent flow are computed explicitly with a resolution near to Kolmogorov scale (no model is required). Consequently, this approach is too expensive to be used in most cases of industrial interest (high Reynolds number and complex geometry). For RANS, the modeling of the whole scales is too dependent on the characteristics of particular flows to be used as a method of general applicability.

The initial applications of LES were the building-block flows such as homogeneous turbulence, mixing layers and plane channel flows. These applications that are geometrically simple and performed usually at low Reynolds numbers are still used to this date to validate new SGS models. This is due to their relative physical simplicity and the availability of high quality and well-documented DNS data about them. Due to the advancement in computer technology and the deeper insight into turbulence physics gained over the years, LES has also demonstrated its capabilities in relatively complex flows such as transitional flows, relaminarizing flows, separated flows, and compressible flows and in research fields such as aero-acoustic and combustion. These types of flows were simulated using LES at Reynolds numbers that could not be reached by DNS.

The LES philosophy is based on the trivial observation that the loss of information incurred during filtering places a limit on the achievable accuracy of LES [1]. Having that in mind, a broad array of SGS models for LES purposes have been developed using a variety of different approximations. Such models have been successful in many flows, yet they all have shortcomings. There are several issues that heavily influence, individually or in combination with others, the performance of any SGS model to achieve that limit imposed by the scale-separation operation:

- Filtering is one of these factors. The interaction between subgrid and resolved scales depends strongly on the choice of the filter. This causes the SGS model to be filter dependent. In LES, the filter width as well as the details of the filter shape are free parameters. This allows the control of the effective resolution of the simulation and to establish the relative importance of the different portions of the resolved spectrum. However, due to the lack of a straightforward and robust filtering procedure for inhomogeneous flows, most LES performed to date have not made use of explicit filtering [2]. Instead, the implicit filtering, which is introduced by the spatial discretization, has been extensively used despite its shortcomings. Indeed, compelling reasons such as consistency and control of numerical errors motivate the adoption of a more systematic approach [3, 4]. The use of explicit filtering gives the ability to control the filter shape and by consequence opens the whole new horizons in LES modeling, since it makes it feasible to look at SGS modeling and filtering independently of numerical issues [5, 6]. Inhomogeneity of most of the flows usually poses problems to the SGS models because they require the use of inhomogeneous filters. Such filters do not in general commute with spatial derivative operators. This introduces new unclosed terms that, in principle, should be modeled. Modeling techniques are not available for these terms though some works have given some insight on how to treat them [2].

- The numerics is another factor. Even when using high-order accurate numerical methods, the numerical discretization errors can far exceed the subgrid term [7]. Using a grid resolution much finer than the filter width to remedy to the problem is rarely done because of the expense. Accounting for discretization errors in the formulation of the SGS model itself is another alternative. It is not generally taken into account in the current SGS models though some types of dissipative numerical discretization have been proposed as implicit SGS models. A purely numerical SGS model for LES known as the monotonically integrated LES (MILES) was proposed [5, 8]. The MILES method is based on the observation that truncation errors due to the discretization of Navier-Stokes equations introduce numerical dissipation with implicit effects qualitatively similar to the effects of the explicit SGS models. A discussion of the shortcomings of the method was provided by Garnier *et al.* [9]. An ameliorated approach was proposed by Domaradzki *et al.* [3]. In the latter, an effective numerical eddy viscosity acting in dissipative numerical schemes used in monotonically integrated LES is computed.

- The SGS modeling is still lacking generality because current SGS models are derived under restrictive assumptions regarding the nature of the subgrid turbulence. In particular, it is generally assumed that the subgrid scales are locally homogeneous and isotropic and that the filter scale is in an inertial range. However, these assumptions are violated in flow situations of common interest such as near wall. Development of wall models that will not require the refinement of the near-wall grid with increasing Reynolds number is considered by many to be the most pressing challenge in LES research [7, 10].

Despite the few shortcomings reported here and there about LES [11-13], it is considered by most of the researchers the future of Computational Fluid Dynamics (CFD). Indeed, this approach can be used to provide reliable data to other new turbulence-tackling methods in dynamical and geometrical circumstances for which DNS is impossible to use. Also, LES can be used directly as a reliable industrial design tool.

2.2. LES Formulation: Filtering

In the standard LES formulation, turbulent motion, defined by the velocity field u, is separated into large scales, defined by $\bar{u}$, and SGS or small scales defined bye u_{SGS}. This is done by spatially filtering the velocity field with a kernel, $G\Delta(x)$, proposed by Leonard [14]:

$$\bar{u}(x) = \int u(x')G_\Delta(x - x')dx' \tag{2.1}$$

The convolution kernel plays the role of a low-pass filter that eliminates scales smaller than the filter width Δ. This results in smoothing the signal u which is usually random and unpredictable.

By applying a spatial filter to the (non-dimensionalzed) isothermal, incompressible, three-dimensional and time-dependent continuity and Navier-Stokes equations:

$$\overline{\frac{\partial u_i}{\partial x_i}} = 0 \tag{2.2}$$

$$\overline{\frac{\partial u_i}{\partial t}} + \overline{\frac{\partial u_i u_j}{\partial x_j}} = -\overline{\frac{\partial p}{\partial x_i}} + \frac{1}{Re_\tau}\overline{\frac{\partial^2 u_i}{\partial x_j \partial x_j}} \tag{2.3}$$

The filtered Navier-Stokes and filtered continuity equations can then be written as:

$$\frac{\partial \bar{u}_i}{\partial x_i} = -M_i \tag{2.4}$$

$$\frac{\partial \bar{u}_i}{\partial t} + \frac{\partial \bar{u}_j \bar{u}_i}{\partial x_j} = -\frac{\partial \bar{p}}{\partial x_i} + \frac{1}{Re_\tau}\frac{\partial^2 \bar{u}_i}{\partial x_j \partial x_j} + \frac{\partial \tau_{ij}}{\partial x_j} - N_i \tag{2.5}$$

where:

$$\tau_{ij} = \bar{u}_i \bar{u}_j - \overline{u_i u_j} \tag{2.6}$$

τ_{ij} is the subgrid stress tensor that must be modeled because it contains terms that cannot be written as functions of the filtered field $\bar{u}$. $Re_\tau = u_\tau . L/v$ is the friction Reynolds number based on a characteristic length L and friction velocity u_τ. v is the fluid kinematic velocity.

For LES of turbulent flows of practical interest that often take place in bounded domains, a local convolution filter is used to adapt the local features of the flow, for instance near the walls. In these cases, commutation errors represented by M_i and N_i arise due to the fact that the filter operation does not commute with differentiation:

$$\overline{\frac{\partial f}{\partial x}} \neq \frac{\partial \bar{f}}{\partial x} \tag{2.7}$$

In a symbolic form it reads:

$$\left[G_*, \frac{\partial}{\partial r}\right] \neq 0., r = t, x \tag{2.8}$$

Consequently M_i and N_i can have the following forms:

$$M_i = \left[G_*, \frac{\partial}{\partial x}\right](u_i) \tag{2.9}$$

$$N_i = \left[G_*, \frac{\partial}{\partial t}\right](u_i) + \left[G_*, \frac{\partial}{\partial x_j}\right](u_i u_j) + \left[G_*, \frac{\partial}{\partial x_i}\right](p) + \frac{1}{Re_\tau}\left[G_*, \frac{\partial^2}{\partial x_i \partial x_j}\right](u_i) \tag{2.10}$$

Though most of the LES-related works have been directed toward modeling the subgrid stress tensor term τ_{ij}, some works have rather focused on the issue of commutation error M_i and N_i [2]. These commutation errors are often overlooked in LES codes. The argument is they are relatively insignificant compared to any other type of errors that can arise from the SGS modeling or the numerical schemes. Some works have been dedicated to this issue, however, in order to eliminate the commutation errors or at least to bound them. This is done by deriving new filtering operators as an alternative for the classical convolution product. Ghosal and Moin [15] defined filters that commute at the second order with the derivation in space (Second Order Commuting Filter, SOCF). This is based on a change of variable that allows the use of a homogeneous filter. Vasilyev *et al.* [16] extended it to a higher order commuting filters.

The most common filters used in LES are the spectral cutoff filter, the Gaussian filter, and the box or top-hat filter (see Table **2.1**). They are called second order filters. By this, it is implied that the difference between the solution u and the filtered solution $\bar{u}$ scales with Δ^2 in case when u is sufficiently smooth and Δ sufficiently small. The spectral cutoff filter cleanly separates between scales. However, when filtering spatially localized phenomena, it causes nonlocal oscillatory behavior. Moreover, because its kernel has negative lobes, the resulting stress tensor does not follow the realizability conditions cited in [17]. The box filter, on the other hand, has good spatial localization but does not allow unambiguous separation between scales because of spectral overlap. The Gaussian filter has intermediate localization properties in both physical and spectral space, although it is closer to the box filter. For Finite Volume (FV) codes, the natural filter introduced by the discretization is Gaussian.

Table 2.1: Filter Kernels in Physical and Spectral Space of the Most Popular Large-Eddy Filters.

Filter	Filter kernel $G(x - \zeta)$	Fourier transform $H(k)$
Top-hat	$\dfrac{1}{\Delta^3}\ if\ |x_i - \zeta_i| < \dfrac{\Delta_i}{2};\ 0\ \text{otherwise}$	$\displaystyle\prod_{i=1}^{3} \dfrac{\sin\left(\frac{k_i \Delta_i}{2}\right)}{k_i \Delta_i / 2}$
Gaussian	$\displaystyle\prod_{i=1}^{3} \left(\dfrac{\alpha}{\pi \Delta_i^2}\right)^{1/2} e^{-\frac{\alpha(x_i - \zeta_i)^2}{\Delta_i^2}}$	$\displaystyle\prod_{i=1}^{3} e^{-\frac{(k_i \Delta_i)^2}{4\alpha}}$
Spectral cut-off	$\displaystyle\prod_{i=1}^{3} \dfrac{\sin\left(\frac{\pi(x_i - \zeta_i)}{\Delta_i}\right)}{\pi(x_i - \zeta_i)}$	$1\ \text{if}\ |k_i \Delta_i| < \pi\ ;\ 0\ othrwise$

The spatial filter that defines the basic LES equations involves the selection of a filter width Δ such that $l_d \ll \Delta \ll L$. This is made in order to be able to restrict the large-eddy modeling to the universal structures in the inertial range [18]. Here L refers to an integral length scale of the flow and l_d is the dissipative length scale. With this choice, large reduction of the computational effort required to simulate the flow is achieved, while the smallest retained structures are much smaller than the integral length scale and hence can be modeled without explicit reference to problem-specific details.

2.3. LES Modeling: Subgrid Stress

SGS models must be robust across many different types of flows including free and wall-bounded turbulence, transition flows and separated flows. They should be, in principle, free of flow-dependent adjustable parameters. They must be capable of operating efficiently and accurately at spatial resolutions substantially less than those required for full DNS and with numerical methods that can be applied to flows with complex boundary conditions. Most importantly, they should provide a framework for extension to mixing flows including the transport of passive and reactive scalars, combusting flows and strongly compressible flows. Several reviews [10, 19] have been dedicated to the different SGS models that have been constantly developed, tested, criticized and eventually modified and ameliorated for use in the large eddy simulations of turbulent flows. Most of these SGS models were developed in physical space, while others in the spectral space. The latter models are extensively covered by Lesieur [20].

Regardless of the nature of the space in which they were developed, commonly used SGS models fall into two general categories: eddy viscosity models, and similarity models. The latter models can be referred to as non-eddy viscosity models. A third category called mixed models can be considered as well but it is nothing than a combination of the two main categories. The eddy viscosity models and the similarity models illustrate two different approaches to SGS modeling. The eddy viscosity models assume that the form of the SGS stress tensor is proportional to the resolved rate-of-strain. They attempt to provide expressions for the proportionality coefficient, the eddy viscosity. In the similarity approach, the primary modeled quantity is the full velocity field. No restrictive form of the SGS stress is assumed and for that reason they are usually called non-eddy viscosity models.

2.3.1. Eddy Viscosity Models

The most common class of SGS models is the eddy viscosity type. These models are generally postulated by analogy to the stresses produced by molecular viscosity. The notion that the effect of the subgrid stresses are increased transport and dissipation (these effects are caused by the molecular viscosity in laminar flows) is at the origin of this class of SGS models [21]. SGS models have been intensively used in LES of turbulent flows owing to their balanced mixture of physical content and mathematical simplicity [10]. Their extension from homogeneous and isotropic turbulence in sheared flows, rotating flows, transitional flows and compressible flows, has however created some problems. In order to extend the range of their applications, they have been constantly the subject of improvements since the first well-known Smagorinsky model.

The most popular eddy-viscosity SGS model is the algebraic model proposed by Smagorinsky [22]:

$$\tau_{ij} - \frac{1}{3}\delta_{ij}\tau_{kk} = -2\nu_{SGS}\overline{S_{ij}}, \tag{2.11}$$

$$\nu_{SGS} = (C_s\Delta)^2\,|\bar{S}| \tag{2.12}$$

where δ_{ij} is the Kornecker Delta taking value 1 when $i = j$ and 0 otherwise. C_s is the Smagorinsky constant. $|\bar{S}| = |2\overline{S_{ij}}\,\overline{S_{ij}}|^{1/2}, \overline{S_{ij}} = \frac{1}{2}(\partial_j\bar{u}_i + \partial_i\bar{u}_j)$ is the resolved rate-of-strain tensor. The constant C_s can be deduced analytically by assuming a filter cutoff in the inertial range. It was estimated to be ≈ 0.18 [21]. For unstructured grids, the filter cutoff length Δ is evaluated as the cube root of the volume of the filtering cell following Deardorff's proposal [23], $\Delta = \Omega^{1/3}$.

For FV codes using second order discretization schemes, the value of the filter width used with unstructured arrangement is usually $\Delta = 2\Omega^{1/3}$ For 3D regular Cartesian meshes, this yields $\Delta = 2h$ where h is the grid spacing. This is motivated by the fact that in such codes any structure smaller or equal to $4h$ is not well discretized [24].

The Smagorinsky model, although relatively successful, is not without problems [3]. For example, to simulate wall-bounded flows, several modifications are required. The value of the parameter C_s in the bulk of the flow has to be reduced from 0.18 to 0.065 [25], which reduce the eddy viscosity by an order of magnitude. In regions close to surfaces, the value has to be reduced even further [10]. This problem arises from the assumption upon which the Smagorinsky model is based. At the subgrid scales, the assumption that production balances dissipation of the turbulent kinetic energy is not accurate [21]. It has been proved that it is violated in the presence of mean shear where the large-scale mean velocity gradient is probably overestimated. Thus, it must be supplemented at the wall by empirical wall functions [26].

One successful recipe is the Van Driest damping [27] that has long been used to reduce the near wall eddy viscosity in RANS models and which is dependent on the distance from the wall, it reads:

$$C_S = C_{S0}\left(1 - e^{-\frac{Y^+}{A^+}}\right)^2, Y^+ = \frac{yu_\tau}{\nu}, u_\tau = \sqrt{\tau_w/\rho_f} \tag{2.13}$$

where Y^+ is the distance from the wall in viscous wall units, u_τ is the shear velocity, τ_w the wall shear stress and A^+ is a constant usually taken to be approximately 25 [25].

Piomelli *et al.* [28] proposed an alternate form:

$$C_S = C_{S0}\left(1 - e^{\left(\frac{-Y^+}{A^+}\right)^3}\right)^{1/2} \tag{2.14}$$

It was demonstrated that this form yields the correct asymptotic behavior of the subgrid viscosity which decreases as Y^{+3} in the near-wall region contrary to the Van Driest function. Although these modifications produce the desired results, it is difficult to justify them in the context of the LES. The SGS model should depend solely on the local properties of the flow and it is difficult to see how the distance from the wall qualifies in this regards [21].

Nicoud *et al.* [29] proposed The WALE model (wall-adapted local eddy-viscosity). It is constructed from an operator based on the traceless symmetric square of the gradient velocity tensor $g_{ij} = \partial \bar{u}_i / \partial x_j$:

$$\bar{s}_{ij} = \frac{1}{2}\left(\bar{g}_{ij}^2 + \bar{g}_{ji}^2\right) - \frac{1}{3}\delta_{ij}\bar{g}_{kk}^2 \tag{2.15}$$

Where $\bar{g}_{ij}^2 = \bar{g}_{ik}\bar{g}_{kj}$. The reasons of the choice behind the (2.15) are the following:

1. Taking into account the effects of the strain rate and the rotation rate of the smallest resolved scales,

2. The need for an operator which vanishes in the near-wall region.

It was demonstrated that these properties are met by the operator (2.15) while operators built on $\bar{S}_{ij}$ as in Smagorinsky model do not retain these characteristics. Additional characteristics of this model are that it only uses local information and is invariant to any change of co-ordinate axes. Scaling considerations led them to propose the following model:

$$\nu_{SGS} = C_w \Delta^2 \frac{\left(\bar{s}_{ij}^d \bar{s}_{ij}^d\right)^{3/2}}{\left(\bar{s}_{ij}\bar{s}_{ij}\right)^{5/2} + \left(\bar{s}_{ij}^d \bar{s}_{ij}^d\right)^{5/4}} \tag{2.16}$$

The constant C_w takes the value 0.1. The major drawback of Smagorinsky and WALE models is their inability to represent correctly, with a single universal constant, different turbulent fields as it was noted by Wong [30].

Germano *et al.* [31] proposed a dynamic procedure for the Smagorinsky model in an attempt to rid it of all its shortcomings cited in [30]. This procedure exploits the spectral information on the energy content of the smallest resolved scales provided by LES calculation to dynamically adjust the Smagorinsky constant as the calculation progresses.

In this dynamic procedure, only the interaction between the subgrid scales and the smallest resolved scales is accounted for in order to dynamically compute the Smagorinsky constant C_s. The calculation is based on the Germano algebraic identity [32] between the resolved turbulent stresses and the subgrid scale stresses obtained using two filters, the grid filter f and the test filter g:

$$\tau_{fg}\left(u_i, u_j\right) = \overline{\tau_f\left(u_i, u_j\right)}^g + \tau_g\left(\bar{u}_i^f, \bar{u}_j^f\right) \tag{2.17}$$

Where τ is defined as:

$$\tau\left(u_i, u_j\right) = \bar{u}_i\bar{u}_j - \overline{u_i u_j} \tag{2.18}$$

The originality of the Germano identity lies in the fact that it links the turbulent stresses calculated at the two filters f and g, to the resolved turbulent stress $\tau_g\left(\bar{u}_i^f, \bar{u}_j^f\right)$. Using the well-known Smagorinsky formulation, the subgrid stresses take the form:

$$\tau_f^a\left(u_i, u_j\right) = -c_s^f l_f^2 \bar{S}^f \bar{S}_{ij}^f \tag{2.19}$$

$$\tau_{fg}^a\left(u_i, u_j\right) = -c_s^g l_g^2 \bar{S}^g \bar{S}_{ij}^g \tag{2.20}$$

where c_s^f and c_s^g are unknown dimensionless parameters. $\bar{S}^{f,g} = \left(2\bar{S}_{ij}^{f,g}\bar{S}_{ij}^{f,g}\right)^{1/2}$ where $\bar{S}_{ij}$ is the large-scale rate tensor. l_f and l_g are the grid filter width and test filter width, respectively. The Germano identity then gives:

$$c_s^g l_g^2 \bar{S}^g \bar{S}_{ij}^g = \overline{c_s^f l_f^2 \bar{S}^f \bar{S}_{ij}^f}^g - \tau_g^a\left(\bar{u}_i^f, \bar{u}_j^f\right) \qquad (2.21)$$

Germano *et al.* [31] used as starting point the following two approximations:

$$c_s^f = c_s^g \ and \ \overline{c_s^f l_f^2 \bar{S}^f \bar{S}_{ij}^f}^g = c_s^f l_f^2 \overline{\bar{S}^f \bar{S}_{ij}^f}^g \qquad (2.22)$$

These two assumptions allow solving for c_s^f using the contracted form of (2.21):

$$c_s^f = \frac{\tau_g\left(\bar{u}_i^f, \bar{u}_j^f\right)\bar{s}_{ij}^f}{2l_f^2 \overline{\bar{s}^f \bar{s}_{ij}^f}^g - 2l_g^2 \bar{s}^g \bar{s}_{ij}^g \bar{s}_{ij}^f} \qquad (2.23)$$

The Germano dynamic procedure with Smagorinsky model has been extensively tested for incompressible flows by Germano *et al.* [31] and Piomelli [26]. For compressible flows, it was tested by Moin *et al* [33]. It was concluded that the Smagorinsky model, with the dynamic procedure, vanishes in laminar flow and at solid boundary. Also, it has the correct asymptotic behavior in the near-wall region without requiring damping or intermittency functions. By intermittency, it is meant alternation of phases of regular and chaotic dynamics in turbulence. It occurs mainly for the high frequencies modes [34]. The model is also capable of accounting for the backscatter, *i.e.* it has the ability of the subgrid scales to add randomness and energy in some cases to the explicit scales.

Other works have been dedicated to highlight the shortcoming of the Germano dynamic procedure and proposed some improvements. Ronchi *et al.* [35] highlighted some of the problems associated with the contracted form used by the Germano dynamic approach. Lilly [36] has used a least squares technique to minimize the difference between the closure assumption and the resolved stresses. This modification proved effective in removing the source of singularity and is believed to extend the method applicability. Ghosal *et al.* [37] used an integral formulation of the Germano identity that rigorously removed the mathematical inconsistency at the expense of having to solve an integral equation at each time step.

A Lagrangian formulation of the dynamic Smagorinsky model has been proposed by Anderson *et al.* [38]. It combines some of the best features of the local and averaged formulations. In order to incorporate averaging in an inhomogeneous, complex geometry, and unsteady flow, the model accumulates the required statistics over fluid-particle trajectories, instead of averaging over directions of statistical homogeneity. Wong [30] proposed a multidimensional statistical-dynamic closure method for the linear and nonlinear anisotropic representation of the SGS stresses. During computations, the SGS representation will be locally and dynamically adjusted to match the statistical structure of the resolved turbulent eddies.

Germano [39] proposed a new formulation of the subgrid scale dynamic model that should force the statistical equivalence of different simulations at different resolution levels. This new formulation is based on the idea that two different LESs are statistically equivalent if they produce the same statistical representation. The basic ingredient of this new formulation is to derive the Smagorinsky coefficient C_s by imposing that two different large eddy representations of a turbulent flow: the computed LES and a tested LES at a higher resolution are equivalent from a statistical point of view. This equivalence is reinforced by requiring that the Reynolds stresses extracted from the different representations are the same. An added advantage of this new dynamic procedure is that the mathematical inconsistency of the older formulations is removed.

In all the works mentioned above, the time scale of the large eddies is often taken equal to $|\bar{S}|^{-1}$, which is strongly affected by the mean flow and by the largest structures. To remedy to this problem, one-equation models have been developed and used as SGS models for large-eddy simulations [37]. For that purpose, a transport equation for the subgrid scale energy k_{SGS} is solved to obtain the velocity scale. Their advantage is a more accurate prescription of the SGS time scale compared to algebraic eddy-viscosity models. However, the expense involved in solving an additional equation does not seem to be justified by improvements in the accuracy [10]. The one-equation model can be written as:

$$v_{SGS} = C_v k_{SGS}^{1/2} \Delta \tag{2.24}$$

$$\frac{\partial k_{SGS}}{\partial t} + \frac{\partial}{\partial x_j}\left(k_{SGS}\bar{u}_j\right) = \frac{\partial}{\partial x_j}\left[\left(C_k k_{SGS}^{\frac{1}{2}}\Delta + v\right)\frac{\partial k_{SGS}}{\partial x_j} + \tau_{ij}\bar{u}_i\right] - C_\epsilon \frac{k_{SGS}^{\frac{3}{2}}}{\Delta} - \tau_{ij}\bar{S}_{ij} \tag{2.25}$$

Ghosal *et al.* [15] discussed the limitations and inconsistencies of the Germano dynamic procedure and proposed a new formulation that rectified them. The new formulation leads to an integral equation whose solution yields the model coefficient as a function of position and time. The formulation is called the dynamic localization model. It uses the SGS kinetic energy equation to derive a velocity scale. This formulation does not admit backscatter. It was tested with success in both decaying isotropic turbulence and flow over backward-facing step. The method can be applied to general inhomogeneous flows. It does not suffer from the mathematical inconsistencies inherent in the dynamic procedure of Germano *et al.* [31].

2.3.2. Scale-Similar and Mixed Models

It was proved by many studies of turbulence phenomena especially those that focused on the nonlinear interactions between different scales that the relation between resolved large scales and the subgrid scale motions is far from being dissipative as postulated by the eddy viscosity models [22]. Indeed, correlations between LES results using Smagorinsky model and DNS results concerning the SGS stress tensor are less than 30% [19]. For that reason a new type of SGS models called scale similar models have been developed.

Similarity models are based on the assumption that the most active subgrid scales are those closer to the cutoff wave number and that the scales with which they interact most are those right above the cutoff. They are believed to reproduce the details of the stresses and the energy exchange accurately on a local level, in particular, the correlation that exists between large scale, energy producing events and energy transfer to and from small scales [19,40].

The scale-similar model can be obtained by applying a second filter $\hat{G}$, with characteristic length $\tilde{\Delta} > \Delta$, to the velocity field. The mixed model can be parameterized as:

$$\tau_{ij} = -2v_{SGS}\bar{S}_{ij} + C_B(\widetilde{\bar{u}_i\bar{u}_j} - \widetilde{\bar{u}}_i\widetilde{\bar{u}}_j) \tag{2.26}$$

C_B is the scale-similar constant that is usually taken equal to 1. It can also be evaluated dynamically in addition to the Smagorinsky coefficient in the so called dynamic mixed model. The two last terms of Equation (2.26) represent the scale-similar model. The eddy viscosity contribution provides the dissipation that is underestimated by the scale-similar part alone as it was first developed by Bardina *et al.* [41]. The mixed model has been tested by Anderson *et al.* [38] for the simulation of homogeneous decaying isotropic turbulence with a Lagrangian dynamic formulation. In this work, averaging is performed over fluid trajectories, which makes the model applicable to complex flows without directions of statistical homogeneity.

The Lagrangian ensemble average is based on the consideration that the memory effects should be calculated in a Lagrangian framework, following the fluid particle, rather than at an Eulerian point, which sees different particles with different histories at each instant. This average is defined as:

$$<f> = \int_{-\infty}^{t} f(t')W(t - t')dt' \tag{2.27}$$

where the integral is carried out following a fluid path-line. W(t) is usually chosen to be an exponential function to give more weight to recent times [42]. This Lagrangian formulation has been tested in forced and decaying isotropic turbulent flows. The results have been compared with those derived from LES using dynamic Smagorinsky and dynamic mixed models. Net amelioration has been gained in term of SGS dissipation of kinetic energy.

Many investigations have investigated a bench of SGS models ranging from the simple Smagorinsky model to the Lagrangian two-parameter dynamic mixed model [4, 40, 43, 44]. They all concluded that models that include a scale similar term and a dissipative one, as well as the Lagrangian ensemble averaging, give results in the best agreement with DNS and experimental data. This is mainly because of their ability to predict non-equilibrium effects.

Some other non-eddy viscosity models have been recently developed [45, 46, 47, 48, 49, 50]. They are based on an approximation of the unfiltered velocity either by de-convolution or interpolation techniques such as inverse modeling based on polynomial inversion and increment SGS model. Domaradzki *et al.* [46] made an estimate of the unfiltered velocity by expanding the resolved large-scale velocity field to subgrid scales two times smaller than the grid scale. The estimation procedure consisted of two steps. The first step used properties of the filtering operation and the presentation of quantities in terms of basic functions such as Fourier polynomials. In the second step, the phases associated with the newly computed smaller scales are adjusted in order to correspond to the small scale phases generated by nonlinear interactions of the large-scale field. The estimated velocity field is expressed entirely in terms of the known, resolved velocity field without any adjustable constants. In all tested cases, the new model performed better than or comparable to classical eddy viscosity models for the majority of physical quantities. In particular, all components of the subgrid scale stress tensor are predicted accurately. This procedure naturally accounts for backscatter without any adverse effects on the numerical stability. It should be noted that the Bardina model [41] that was at the heart of the development of similarity models, is obtained by formally approximating the full velocity field u_i by the filtered velocity $\bar{u}_i$ and then computing τ_{ij} using Equation (2.6).

Most of the commercial and industrial CFD codes that are in use nowadays were initially developed based on RANS. When they have been extended to perform LES, the obvious choice was the use of the eddy viscosity models (Smagorinsky and Germano-Lilly dynamic models) to account for the SGS effects in a similar way to the eddy viscosity models for RANS. This is mainly for coding convenience.

2.4 LES: Numerical Issues

There are many important issues in LES which make the validation of the results, and of the SGS models themselves, essential. The most obvious ones are those related to the physical modeling of the subgrid stresses, which is in many cases based on engineering approximations, justified by little more than dimensional considerations [21]. Numerical issues are also important, since most models use as input the velocity gradients of the filtered field, which depend on the smallest scales resolved by the simulations, and therefore strongly influenced by numerical errors. Consistency with filtering of the physical models and numerical schemes adopted is as important as the previous issues. So, it is clear that filtering, modeling, and numerical techniques are interrelated but in which way is still unclear [51].

Conserving turbulent kinetic energy is important in LES to ensure that the real performance of the SGS model employed is not masked by numerical errors. This is an important issue in particular in the near-wall region or in transitional regions of the flow. Using high order numerical schemes is a sufficient condition to minimize such errors. However, their use is not straightforward for unstructured arrangements that are the main feature in CFD codes aimed at turbulent flows in complex geometries. Nonetheless, a careful discretization of the momentum and mass equations in space and time can give, even with a low order scheme, good conservation properties [52].

Conserving the momentum is natural when using, with a finite volume technique, the conservative form of Navier-Stokes equations. The time advancing scheme plays a major role in the conservation of global kinetic energy. It is also recommended in LES to use high order time schemes at least second order [21].

REFERENCES

[1]   Langford JA, Moser R.D. Optimal LES formulations for isotropic turbulence. J Fluid Mech 1999; 398; 321-346.
[2]   Geurts BJ. Elements of direct and large-eddy simulation. Edwards editions 2003.
[3]   Domaradzki JA, Xiao Z, Smolarkiewicz PK. Effective eddy viscosities in implicit large eddy simulations of turbulent flows. Phys Fluids 2003; A(15)12.
[4]   Piomelli U, Moin P, Ferziger JH. Model consistency in large eddy simulation of turbulent channel flows. Phys Fluids 1988; 31(7).
[5]   Boris JP, Grinstein FF, Oran ES, Kolbe RL. New insights into large-eddy simulation. Fluid Dyn Res 1993; 10; 199.
[6]   Lund TS, Kaltenbach HJ. Experiments with explicit filtering for LES using a finite-difference method. In Annual Research Briefs, Center for Turbulence Research, NASA Ames/Stanford University1997; 91-104 pp.
[7]   Rogallo RS, Moin P. Numerical simulation of turbulent flows. Annu Rev Fluid Mech 1984; 16; 99-137.

[8]   Grinstein FF, Fureby C. Recent progress on MILES for high Reynolds number flows. J Fluids Eng 2002; 124; 848.

[9]   Garnier E, Mossi M, Sagaut P, Comte P, Deville M. On the use of shock-capturing schemes for larde eddy simulation J Comput Phys 1999; 153, 273.

[10]   Piomelli U. Large-eddy simulation: achievements and challenges. Progress in Aerospace Sciences 1999; 35; 335-362.

[11]   Pope S.B. Ten questions concerning the large-eddy simulation of turbulent flows. N Phys 2004; 6; 35.

[12]   Lesieur M. Turbulence in fluids. Kluwer Academic Publishers 1990.

[13]   Ghosal S. Mathematical and physical constraints on large eddy simulation of turbulence. AIAA J 1999; 37(4); 425-433

[14]   Leonard A. Energy cascade in large-eddy simulation of turbulent fluid flows. Adv Geophys 1972; 18; 237.

[15]   Ghosal S, Moin, P. The basic equations for the large eddy simulation of turbulent flows in complex geometry. J Comput Phys 1995; 118; 24-37

[16]   Vasilyev O, Lund TS, Moin P. A general class of commutative filters for LES in complex geometries. J Comput Phys 1998; 146; 82-104.

[17]   Vreman AW, Geurts BJ, Kuerten JGM. Realizability conditions for the turbulent stress in large eddy simulation. J Fluid Mech 1994; 278; 351.

[18]   Sagaut P. Large eddy simulation for incompressible flow. Springer Verlag 2001.

[19]   Meneveau C, Katz J. Scale-invariance and turbulence models for large eddy simulation Annu Rev Fluid Mech 2000; 32; 1-30.

[20]   Lesieur M, Metais O. New trends in large eddy simulations of turbulence. Annu Rev Fluid Mech 1996; 28; 45-82.

[21]   Ferziger JH, Peric M. Computational methods for fluid dynamics. Springer-Verla 1996.

[22]   Smagorinsky, J. General circulation experiments with the primitive equations. Mon Weather Rev 1963; 91; 99.

[23]   Deardorff JW. A numerical study of three-dimensional turbulent channel flow at large Reynolds numbers. J Fluid Mech 1970; 41; 453-465.

[24]   Pope S.B. Turbulent flows. Cambridge University Press 2000.

[25]   Piomelli U, Ferziger JH, Moin P. New approximate boundary conditions for large eddy simulations of wall-bounded flows. Phys Fluids 1989 A(1)6.

[26]   Piomelli U. High Reynolds number calculations using the dynamic subgrid scale stress model. Phys. Fluids 1993; A5 (6).

[27]   Van Driest ER. On the turbulent flow near a wall. J Aero Sci 1956; 23; 1007.

[28]   Piomelli U, Zang TA, Speziale CG, Hissaini MY. On the large eddy simulation of transitional wall-bounded flows. Phys Fluids 1990; A2(2); 257-265

[29]   Nicoud F, Ducros F. Subgrid scale stress modeling based on the square of the velocity gradient tensor. Flow Turbulence Combust 1999; 62; 183-200.

[30]   Wong VC. A proposed statistical-dynamic closure method for the linear or nonlinear subgrid scale stresses. Phys Fluids 1992; A(4)5.

[31]   Germano M, Piomelli U, Moin P, Cabot WH. A dynamic subgrid scale eddy viscosity model. Phys Fluids 1991; A(3)7.

[32]   Germano M. Turbulence: the filtering approach. J Fluid Mech 1986; 238; 325-336.

[33]   Moin P, Squires K, Cabot WH, Lee S. A dynamic subgrid scale model for compressible turbulence and scalar transport. Phys Fluids 1991; A3; 2746.

[34]   Bini M, Jones WP. Particle acceleration in turbulent flows: A class of nonlinear stochastic models for intermittency. Phys Fluids 2007; 19(035104).

[35]   Ronchi C, Ypma M, Canuto VM. On the application of the Germano identity to subgrid scale modeling. Phys Fluids 1992; A(4)12.

[36]   Lilly D.K. A proposed modification of the Germano subgrid scale closure method. Phys Fluids 1992; A(4)3.

[37]   Ghosal S, Lund TS, Moin, P, Akselvoll K. A dynamic localization model for large eddy simulation of turbulent flows. J Fluid Mech 1995; 286, 229.

[38]   Anderson R, Meneveau C. Effects of the similarity model on finite-difference LES of isotropic turbulence using a Lagrangian dynamic mixed model. Flow Turbulence Combust 1999; 62; 201-225.

[39]   Germano M. A statistical formulation of the dynamic model. Phys Fluids 1996 A(8)2.

[40]   Sarghini F, Piomelli U, Balaras E. Scale-similar models for large-eddy simulations. Phys Fluids 1999; A(11)6.

[41]   Bardina J, Ferziger JH, Reynolds WC. Improved turbulence models based on LES of homogeneous incompressible turbulent flows. Department of Mechanical Engineering. Report No. TF-19, Stanford 1984.

[42]   Mohseni K, Kosovic B, Shkoller S, Mardsen JE. Numerical simulations of the Lagrangian averaged Navier-Stokes equations for homogeneous isotropic turbulence. Phys Fluids 2003 A15(2).

[43]   Salvetti MV, Beux F. The effect of the numerical scheme on the subgrid scale term in large eddy simulation. Phys Fluids 1998; A(10)11.

[44]   Zang Y, Street RL, Koseff JR. A dynamic mixed subgrid scale model and its application to turbulent recirculating flows. Phys Fluids 1993; A6(12).

[45]   Brun C, Friedrich R. The spatial velocity increment as a tool for SGS modeling. Modern simulation strategies for turbulent flows. Ed: B.J Geurts, R.T, Edwards Inc. Philadelphia 2001; 57.

[46]   Domaradzki JA, Saiki EM. (1997). A subgrid scale model based on the estimation of unresolved scales of turbulence. Phys Fluids A(9)7.

[47]   Geurts BJ. Inverse modeling for large eddy simulation. Phys Fluids 1997; 9; 3585

[48]   Kerr RM, Domaradzki JA, Barbier G. Small-scale properties of nonlinear interactions and subgrid scale energy transfer in isotropic turbulence. Phys Fluids 1996; 8; 197.

[49]   Scotti A, Meneveau C. Fractal model for course-grained nonlinear partial differential equations Phys Rev Lett 1997; 78; 867.

[50]   Shah K.B., Ferziger J.H. (). A new non-eddy viscosity subgrid scale model and its application to channel flow. In Annual Research Briefs (Center for Turbulence Research, NASA Ames-Stanford University, 1995; 73 p)

[51]   Jiminez J. Data base for the validation of LES computations in transition and turbulence. AGARD R-345. 1993.

[52]   Benhamadouche S. Large eddy simulation with the unstructured collocated arrangement. PhD. thesis, University of Manchester 2006.

CHAPTER 3

Large Eddy Simulation of Particle-Laden Flows

Abstract: In this chapter, the importance of the dispersed two-phase turbulent flows for industrial and environmental applications is highlighted. Both Eulerian and Lagrangian descriptions of particle-laden flows are reviewed. Since the book's focus is on the stochastic modeling for particle transport by subfilter motion in a Lagrangian framework, more emphasis is put on the Lagrangian description. Though the stochastic modeling is so far only tested in a context of one-way coupling between dispersed and carrier phases, the importance that subfilter motion may have in predicting turbulence modulation is also explained.

Success in simulating particle-laden turbulent flows relies heavily on a greater understanding of the interaction of the two phases. This can lead undoubtedly to increases in performance, reduction in cost and/or improved safety in systems where they are encountered. It also increases the quality of predictions of the effectiveness of natural flow phenomena for dispersing particulate pollutants to acceptable concentration levels.

In principle, the direct numerical simulation (DNS) of turbulent flows, involving a large number of particles, with appropriate boundary and initial conditions would describe completely the two-phase flows. Due to the high computational cost of DNS, both the velocity field of the carrier phase and trajectories of particles can be calculated through Large Eddy Simulations (LES). Yet another method, Stochastic Modeling (SM) coupled to RANS calculations can be used. The aim of RANS/SM is to reduce the computational effort through generating a synthetic turbulent flow field with statistical properties of interest identical to that of the real turbulent flow.

Keywords: Large eddy simulation; multiphase flows; turbulent flows; Lagrangian description; turbulence modulation; particle equation; drag laws; stochastic process; dispersion; deposition.

3.1. INTRODUCTION

In this chapter, the importance of the dispersed two-phase turbulent flows for industrial and environmental applications is highlighted. Both Eulerian and Lagrangian descriptions of particle-laden flows are reviewed. Since the book's focus is on the stochastic modeling for particle transport by subfilter motion in a Lagrangian framework, more emphasis is put on the Lagrangian description. Though the stochastic modeling is so far only tested in a context of one-way coupling between dispersed and carrier phases, the importance that subfilter motion may have in predicting turbulence modulation is also explained. We start with the definition of a particle-laden turbulent flows and the importance of its prediction.

The flow of fluids laden with particles, droplets and/or bubbles or the so-called dispersed flow is a subcategory of multi-component, multiphase flow that covers a wide spectrum of flow conditions and applications. Multiphase flows can be subdivided into four categories; gas-solid, gas-liquid, liquid-solid and three-phase flows. As a subcategory of two-phase flows, dispersed two-phase flows occur when one phase (called the continuous phase) is a continuum and the other phase (called the dispersed phase) appears as separate inclusions dispersed within the continuous one.

The word particle refers to either solid particle, droplet of liquid, or a gaseous bubble and it is defined as a self-contained body with usual dimensions between about 0.5 μm and 10 cm [1]. These inclusions are separated from the surrounding medium by a recognizable interface. When the dispersed phase is characterized by a distribution in size, one speaks of a polydispersed two-phase flows. Particle-laden flows can be broadly classified in three categories determined on the basis of inter-particle collisions: collision-free flow (dilute flow), collision-dominated flow (medium concentration flow) and contact-dominated flow (dense flow) [2]. Inertial particles dispersion in a gas stream is a type of two-phase flows that probably characterizes the greatest number of cases occurring in industry and nature.

The importance of particle dispersion and deposition in two-phase flows has been well recognized in many different fields of research and industry. Indeed, the dispersion of one phase within another one increases considerably the area of the separating interface and thus allows better mass and energy transfer between the two phases. These higher transfer rates explain why the dispersed regime is preferable [3]. In most applications one is interested in how particles are transported by turbulent flows, how they modulate the turbulence and where they eventually deposit.

Success in simulating particle-laden turbulent flows relies heavily on a greater understanding of the interaction of the two phases. This can lead undoubtedly to increases in performance, reduction in cost and/or improved safety in systems where they are encountered. It also increases the quality of predictions of the effectiveness of natural flow phenomena for dispersing particulate pollutants to acceptable concentration levels.

In principle, the direct numerical simulation (DNS) of turbulent flows, involving a large number of particles, with appropriate boundary and initial conditions would describe completely the two-phase flows. Due to the high computational cost of DNS, both the velocity field of the carrier phase and trajectories of particles can be calculated through Large Eddy Simulations (LES). Yet another method, Stochastic Modeling (SM) coupled to RANS calculations can be used. The aim of RANS/SM is to reduce the computational effort through generating a synthetic turbulent flow field with statistical properties of interest identical to that of the real turbulent flow.

3.2 Analytical Description of Dispersed Two-Phase Turbulent Flows

In particle-laden flows, the carrier phase is governed by the conservation of mass and momentum equations in the Eulerian framework. The motion of the dispersed phase occurs due to forces generated by the moving fluid and acting through the interfaces, such as fluid drags force, as well as other forces such as gravity. All these forces with Newton's second law govern the trajectory of motion of each particle in the fluid flow domain. The Eulerian equations for the carrier phase along with the Lagrangian or Eulerian equations for the particle phase are referred to as the 'first-principle' equations in most of the reviews dedicated to dispersed two-phase turbulent flows [2-8].

To tackle dispersed two-phase turbulent flows, two analytical descriptions are used: the hybrid Eulerian-Lagrangian approach, or simply the Lagrangian approach and the Eulerian-Eulerian approach, or simply the Eulerian approach. This classification is made according to the reference frame in which the dispersed phase is treated since in both approaches, the carrier phase is treated in an Eulerian coordinate system.

In the Lagrangian reference frame, individual particles or clouds of particles are treated in a discrete way. The reference frame moves with the particles, and the instantaneous location of each particle is determined by reference to its origin and the time elapsed. Lagrangian methods employed for particle tracking are conventionally based on the equation of motion for spherical particles in high-Reynolds number turbulent flows [9]. The dispersed phase is assumed to be heavy and smaller than the Kolmogorov micro-scales. As a prerequisite computational sequence, the flow field has to be known since tracking individual particles directly relies on its properties, *i.e.* velocity field and turbulence statistics. Thus, the parameterization of particle dispersion is intimately tied to the dynamics of the turbulence field. This is the case for dispersed phase smaller than the Kolmogorov micro-scales, whose interaction with turbulence is commonly termed one-way coupling by reference to the weak effect of particle momentum on turbulence.

An important issue in Lagrangian treatment of the dispersed phase is the number of particles tracked. Since each particle is considered as one realization, it is required that the number of particles N_p be large enough to provide accurate statistics generated *via* ensemble averaging over the number of particles. It is shown that the statistical sampling error decreases as $N_p^{0.5}$ [2]. The Lagrangian approach becomes computationally inefficient for the prediction of situations where the required number of particles is so large that the Lagrangian simulation becomes too costly. As an alternative and like the carrier phase, the dispersed phase can be treated as a continuous phase in an Eulerian framework.

The Eulerian approach for simulating turbulent dispersion has its own advantages as compared to Lagrangian methods. For flow laden with a large amount of particles the quantitative description of the variation in particle concentration is much simpler by means of the Eulerian method while statistical sampling is required with the Lagrangian description. The Eulerian approach allows both phases to be computed over a single grid whereas the Lagrangian methods require the interpolation of quantities between the fixed grid nodes and the local position of particles. Lagrangian methods may also face problems whenever the cloud of particles tracked is larger than the fluid parcel over which volume averaging is performed [7].

The Eulerian approach is computationally efficient but the price paid for this is that assumptions must be introduced in the formulation of the dispersion tensor. Conversely, in the Lagrangian approach, trajectory realizations are explicitly simulated by the computer and subsequent averaging is carried out. Therefore, the range of applications is dramatically increased but the price that has to be paid is more time-consuming runs [2]. Also, treating particles *via* the Lagrangian formalism is in essence natural because their motion is tracked as they move through the flow field. This preserves their actual non-continuum behavior and accounts for their history effects in a natural way. In addition, if attention is redirected toward turbulence modeling, the Lagrangian approach holds a fundamental advantage over the Eulerian one in the sense that, in general, it does not require closure assumptions for turbulence correlations for particle concentration and velocity fluctuations [2].

3.3 Turbulence Modulation

In general, particles are driven by carrier phase and as a consequence this phase experiences reaction forces from these particles. The contaminating character of particles locally modifies the turbulence, since inside the particle there is no flow field and no eddies. This kind of local modification is expected to be negligible if the particle diameter is much smaller than the Kolmogorov scales [10]. However, when higher particle loadings are considered (volume fraction $\alpha_p > 10^{-6}$) [11], particles significantly affect the turbulent flow field and the two-way coupling between the two phases has to be taken into account, complicating the simulations considerably [11]. At even higher particle concentrations (volume fraction $\alpha_p > 10^{-3}$) particle-particle interaction becomes important. This is a four-way coupling problem [2] (see Fig. **3.1**).

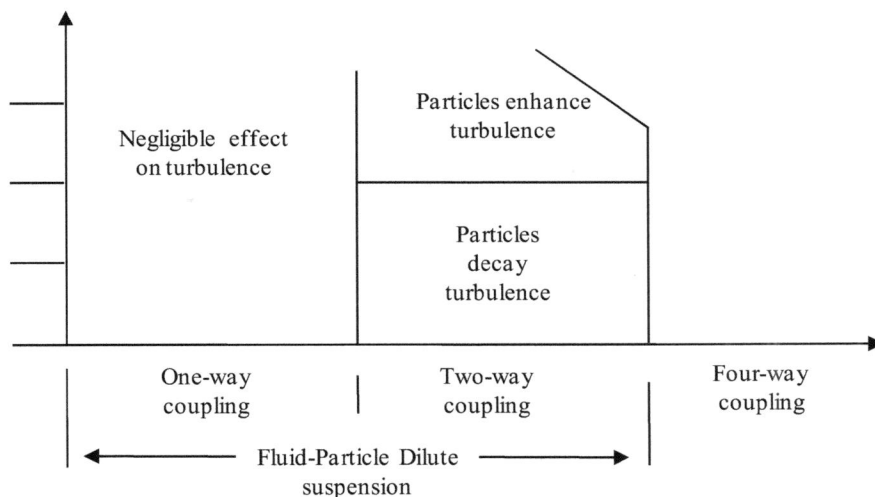

Figure 3.1: Proposed map for particle-turbulence modulation. From Crowe et al. [25]. τ_p and τ_e are the time scales associated with the dispersed and carrier phases respectively.

The presence of particles in a turbulent flow can modulate the turbulence in several ways. Particles can distort streamlines or modify velocity gradients leading to a change in turbulence generation. Large particles can generate wakes yielding a turbulence enhancement while small particles causes a turbulence damping by the drag forces on them since kinetic energy is converted into heat. This has been the general trend shown by the experimental and DNS data. By compiling data up to the year 2000, Crowe [12] stated that the turbulence modulation is correlated with the ratio of the particle size to the Kolmogorov length scale R_{pt}. For $R_{pt} < 1$, the turbulence intensity is attenuated while an increase in turbulence level is observed for $R_{pt} > 1$ as it was noticed also by Gore and Crowe [13]. By investigating experimentally turbulence modulation in gas-solid, gas-liquid and solid-liquid pipe flows, Hosokawa and Tomiyami [14] confirmed that the turbulence modification is correlated with another ratio, Φ. It is the ratio of the eddy viscosity induced by the dispersed phase to the shear-induced viscosity. They showed that turbulence modulation is better correlated with Φ than with R_{pt}. Hadinoto *et al.* [15] indicated that turbulence modulation is also Reynolds number dependent. They experimentally studied a gas-particle turbulent pipe flow with a constant mass loading and at different Reynolds numbers. They concluded that the demarcation criteria proposed by Gore and Crowe [13] for suppression or enhancement of gas turbulence in the presence of particles is unable to explain their findings and other previously published experimental results in particular the work of Maeda *et al.* [16].

It is well known that resolving the flow around thousands to millions of particles to get an accurate particle/turbulence interaction is prohibited by the number of grid points required in a local spherical grid surrounding each particle. Thus, physical models are needed to predict turbulence modulation by the presence of solid particles. Eaton and Segura [17] showed that one million grid points have to be used in a local spherical grid of diameter equal to $25d_p$ to keep errors under 1%.

A commonly applied model is the point-force model that was first used by Squires and Eaton [18]. This model has been improved by Lonholt *et al.* [19] and Sundaram and Collins [20]. In all these models, it was assumed that the particles are much smaller than both the Kolmogorov scale and the grid spacing [21].

As stated by Eaton and Segura [17], the point-force model, when added into a DNS, is believed to capture the effects of particles on the energy-containing eddies while being incapable of capturing the extra viscous dissipation associated with the particle motion. This fact may be exacerbated when the approach is used in the context of LES when the subgrid scale motion is discarded by the filtering operation. Segura [22] performed a well-resolved LES of a channel flow using the point-force coupling model. The single-phase flow and particle motions for light loading cases (mass loading 20%) were accurately predicted compared to the experimental results of Paris and Eaton [23] whereas the turbulence attenuation was grossly under-predicted.

They are many reasons behind the failure of the point force scheme to capture the high levels of local dissipation around the particles. One reason is that most of the Lagrangian models relate the particle drag to the undisturbed fluid velocity which is not available in a two-way coupled LES. The undisturbed fluid velocity field is defined as the velocity field that would exist if the particle was not present [3]. Another reason is the under-prediction of particle drag by Lagrangian tracking models because the models do not account for the effects of small scale turbulence which is missing in the LES field. Thus, the representation of the extra dissipation that occurs at subgrid scale may bring about amelioration of the point-force LES predictions for Lagrangian two-phase flows. Elghobashi [24] also pointed out that in case of a significant two-way coupling, the subgrid scale turbulence model might need modification. This would be specifically important if the particulate phase couples strongly with the small scales turbulence. Whether this is the situation or not should depend upon the matching between the particle and the subgrid scale turbulence time scales.

3.4 Lagrangian Equation of Motion for Dispersed Phase

For the Lagrangian approach, an equation of motion is needed to track the dispersed phase. To solve it, this Lagrangian equation is supplemented by the Eulerian velocity field of the carrier phase interpolated at the particle positions. Subsequently, construction of trajectories and averaging over a certain number of realizations can be performed.

Loosely speaking, the exact equations of motion of the discrete particles are the equivalent of the fundamental equations for the fluid phase that are the Navier-Stokes equations. Yet, the situation is more complex: the equations of motion for discrete particles in a turbulent flow field do not have the same level of validity as the Navier-Stokes equations [3]. The exact form remains a subject of current research.

The calculation of forces acting on various particles within the flow domain depends on the knowledge of the turbulent flow field. Then the new location of all particles at time $(t + \Delta t)$ can be obtained from Newton's second law.

Under some circumstances when the particle size is smaller than the smallest length scales of the flow, particles can be considered as 'point' particles. Thus, instead of computing the forces by solving flow around them, various existing methodologies use expressions for these forces which are obtained from either experimental, analytical or DNS studies on a single or isolated particle. Certainly, the obtained expressions do not imitate exactly the real forces acting on the particle surrounded by other particles. Improvement of these expressions is of more importance for the prediction of medium concentration and dense two phase flows where the inter-particle collisions and two-coupling mechanisms are important [2].

Despite the shortcoming in accurate description of forces, the Lagrangian equations for point particle with expressions for the forces on an isolated particle have been considered as the main starting point in almost all practical applications thus far. They are considered as first-principle equations for the dispersed phase. The Lagrangian equations governing the position x_p and the velocity u_p, along the trajectory of spherical particle, of diameter d_p, and mass m_p in the carrier flow field can be written as:

$$\frac{dx_{p,i}}{dt} = u_{p,i} \tag{3.1}$$

$$m_p \frac{du_{p,i}}{dt} = F_i \tag{3.2}$$

$$\frac{F_i}{m_p} = f_i = f(\rho_p, \rho_f, \mu, d_p, u_{p,i}, g_i, Du_{f,i}/Dt) \tag{3.3}$$

where F denotes the summation of all the forces acting on the particles. ρ_p and ρ_f are the density of the particle material and of the surrounding fluid, respectively. μ is the dynamic viscosity of the surrounding fluid. u_p is the velocity vector of the particle, u_s is the fluid velocity seen, g and $Du_{f,i}/Dt$ are the gravity vector and the acceleration vector of the fluid. The fluid velocity seen u_s is the fluid velocity sampled along the particle trajectory $x_p(t)$, not to be confused with the fluid velocity u_f computed in the Eulerian framework, which is usually known at the centre of the cell or on its nodes.

The problem of predicting these forces on a particle moving in a viscous fluid has been studied in the past 160 years since Stokes first worked out the resistive drag force on a single sphere in 1851 in creeping flow conditions. Basset (1888), Boussinesq (1903) and Ossen (1927) [2] then extended the Stokes's work to the case of a particle accelerating in a fluid at rest. Tchen [25], in 1947, was the first to generalize the investigation for a particle in a turbulent flow. Since then, various contributors have addressed the question and have proposed modifications of the expressions of the forces, however often in *an ad hoc* manner. The derivation of the forces was analytically carried out from first principles by Maxey and Riley (1983) [9].

Maxey and Riley derived the BBO equation (Basset, Boussinesq and Ossen) from first principles for low particle Reynolds numbers ($Re_p = d_p|u_f - u_p| / \nu < 1$). The application of this equation is also limited to particle diameters much less than the local turbulence length scale [10]. Maxey and Riley's equation (MR) is a more complete derivation taking proper account of the effect of spatial variation of the flow field and accounts for the influence of velocity profile curvatures. The final (MR) equation of motion in non-uniform flow is:

$$m_p \frac{du_{p,i}}{dt} = m_f \frac{Du_{f,i}}{dt}|_{x_{p,i}(t)} - 6\pi d_p \mu \left[\mu_{p,i} - \mu_{f,i}(x_{p,i}(t),t) - \frac{d_p^2}{6} \frac{\partial^2 u_{f,i}}{\partial x_j \partial x_j}|_{x_{p,i}(t)} \right]$$

$$+ (m_p - m_f)g_i - \frac{m_f}{2} \frac{d}{dt} \left[u_{p,i} - u_{f,i}(x_{p,i}(t),t) - \frac{d_p^2}{10} \frac{\partial^2 u_{f,i}}{\partial x_j \partial x_j}|_{x_{p,i}(t)} \right]$$

$$- 6d_p^2 (\pi\mu\rho_f)^{0.5} \int_0^t \frac{\frac{d}{dt}\left[u_{p,i} - u_{f,i}(x_{p,i}(t),t) - \frac{d_p^2}{6} \frac{\partial^2 u_{f,i}}{\partial x_j \partial x_j}|_{x_{p,i}(t)} \right]}{(t-\tau)^{0.5}} d\tau. \tag{3.4}$$

The five terms on the right-hand side of Equation (3.4) represent fluid acceleration force, drag force, buoyancy force, added mass force, and Basset history force, respectively. The expression for drag, added mass and Basset history forces are modified due to the flow curvature effect represented by the term containing $\frac{\partial^2 u_{f,i}}{\partial x_j \partial x_j}|_{x_{p,i}(t)}$

In the Lagrangian approach, there is no difficulty in implementing any equation of motion because such a change of equation only requires modifying a few program statements. We may then easily handle more complicated equations than the MR equation. This is actually required in many situations because this equation is limited to the case when particle Reynolds number is small enough. For large particulate Reynolds numbers, a modified MR equation is used. Indeed, for higher Re_p, when exact calculations cannot be performed anymore, it is assumed that the different forces already isolated are still present, but with a generalized expression [3]. Thus, modifications were introduced through empirical correcting factors C_D, C_A, C_H for the drag term, added mass term and the Basset term, respectively. The modified MR equation accounting for higher particulate Reynolds number reads as [6]:

$$\rho_p \frac{\pi d_p^3}{6}\frac{du_{p,i}}{dt} = \frac{\pi d_p^3}{6}(\rho_p - \rho_f)g_i + \rho_f \frac{\pi d_p^3}{6}\frac{Du_{f,i}}{dt} - \frac{\pi d_p^2}{8}\rho_f C_D (u_{p,i} - u_{f,i})|u_p - u_f|$$

$$-\rho_f \frac{\pi d_p^3}{6}C_A \frac{d}{dt}(u_{p,i} - u_{f,i}) - \frac{\pi d_p^2}{4}C_H \frac{(\rho f \mu)^{0.5}}{\pi}\int_0^t \frac{\frac{d}{dt}(u_{p,i}-u_{f,i})}{(t-\tau)^{0.5}}d\tau, \tag{3.5}$$

in which:

$$C_{D=}\frac{24}{Re_p}\left(1 + 0.15Re_p^{0.687}\right), with\ Re_p < 200,$$

$$C_A = 1.05 - 0.0066/A_C^2 + 0.12,$$

$$C_H = 2.86 - 3.12/(A_C^2 + 1)^3,$$

$$A_C = \frac{|u_p - u_f|^2}{d_p \frac{d(u_p - u_f)}{dt}}, with\ A_C < 60. \tag{3.6}$$

For the case of turbulent flows laden with inertial particles where $\rho_p \gg \rho_f$, the drag and gravity forces are the dominant forces [2]. The particle equation of motion can be reduced to:

$$\frac{du_p}{dt} = \frac{1}{\tau_p}\left(u_f - u_p\right) + g. \tag{3.7}$$

Actually, the fact that the added-mass and Basset forces can be neglected with respect to the drag force, when $\rho_p \gg \rho_f$, is not so obvious from the expressions of the different forces. Indeed the particle density is present only with gravity in the expressions of these forces in equation (3.5). This can be justified by going back to the analysis of the basic equations using some dimensional analysis [3]. In Eq. (3.7), the drag force has been written in such a form to bring out the particle relaxation time scale:

$$\tau_p = \frac{\rho_p}{\rho_f}\frac{4d_p}{3C_D|u_f - u_p|} \tag{3.8}$$

The response time of a solid particle or droplet to changes in flow velocity is important to establish non-dimensional parameters to characterize the flow. The response time relates to the time required for a solid particle to respond to a change in fluid velocity field. It now appears as the only scale when it comes to the particle motion and it is most sensitive to particle diameter [10].

In Equation (3.8), the drag coefficient C_D can be replaced by a more general expression valid for higher values of Re_p characterizing regimes beyond the Stokes flow regime. Different expressions for the drag coefficient are proposed for a quite wide range of relative particulate Reynolds number such as [1, 3]:

$$C_D = \frac{24}{Re_p}\left(1 + 0.15Re_p^{0.687}\right) + \frac{0.42}{1 + 4.25 \times 10^4 Re_p^{-1.16}}, with\ Re_p < 3 \times 10^5$$

$$C_D = \frac{24}{Re_p}\left(1 + 0.15Re_p^{0.687}\right), with\ Re_p < 800,$$

$$C_D = \frac{24}{Re_p}\left(1 + \frac{Re_p^{\frac{2}{3}}}{6}\right), with\ Re_p < 1000,$$

$$C_D = \frac{24}{Re_p}\left(0.0183Re_p\right), with\ 1000 < Re_p < 3 \times 10^5. \tag{3.9}$$

In some cases, it might also be necessary to add the lift force to the equation of motion of particles. Saffman (1968) [26] showed that a spherical particle in a shear flow experiences a lift force. The Saffman lift force F_s is given by:

$$F_s = 6.44d_p^2(\mu.\rho f)^{0.5}|\nabla \times u_f|^{-0.5}\left[(u_f - u_p) \times (\nabla \times u_f)\right]. \tag{3.10}$$

It is suggested that comparison between lift velocity, $V_L = 0.3d_p(u_f - u_p) \times (\nabla \times u_f)^{0.5}/v^{0.5}$ [27], and particle dispersion velocity can be a criterion to determine whether or not lift forces may be considered [2]. If the lift drift velocity is much smaller than the dispersion velocity then lift forces may presumably be neglected. For instance, In Arnason's experiments [28], lift velocity and dispersion velocity were estimated to be 1 and 50 cm/s, respectively. Adding a lift force to the equation of motion of particles is not, however, an entirely safe procedure [6]. First, for high particulate Reynolds numbers, assumption of creeping flow around the particle is no longer valid. Thus the problem to be solved is no longer linear and the principle of superposition of solutions is not valid. Also, assumptions underlying the work of Saffman are usually not satisfied. In particular, this work assumes steadiness which is never instantaneously satisfied in turbulence [2].

3.5 Lagrangian LES Description of Particle-Laden Turbulent Flow

Traditional methods for simulation and modeling carried out to investigate particle-laden turbulent flows have demonstrated the shortcomings of RANS methods for the prediction of particle-laden turbulent flows. These shortcomings are related to deficiencies associated with the model used to predict the properties of the Eulerian turbulence field. Accurate prediction of particle transport is strongly dependent upon providing a realistic description of the time-dependent, three-dimensional velocity field encountered along particle trajectories.

Synthetic turbulence built upon these RANS-provided turbulent properties can be generated to match as closely as possible a real turbulence provided that most of the underlying turbulence physics is captured. This is often not the case for turbulent flows involving complex physics taking place in complex geometry. Nonetheless, stochastic construction of synthetic turbulence based on RANS has emerged as a low-cost tool to predict particle-laden turbulent flows. On the other hand, DNS, while an extremely powerful tool for supplying information about particle-turbulence interaction which cannot be obtained from experiment, is not practical for use as a predictive tool. It remains restricted to relatively low Reynolds number turbulent flows as a result of its high computational cost.

An approach which is not as severely restricted in the range of Reynolds numbers as DNS is LES. Despite the high computational cost related to LES compared to the RANS/Stochastic Modeling (SM), it has a significant advantage that it permits a much more accurate and natural accounting of particle-turbulence interactions at an affordable computing cost.

Lagrangian LES and RANS/SM are two practical approaches to investigate particle-laden turbulent flows in the Lagrangian framework and this at both ends of the CFD spectrum. No approach can compete with LES in term of accuracy at the high-cost end whereas RANS/SM claims to provide the best possible results in term of accuracy at the low-cost end. The PDF approach which is a new modern working horse in the area of two-phase flow is definitely costlier than the RANS/SM and LES but its accuracy has not yet been proved lower or higher than LES since it is still an emerging method. Lagrangian LES and RANS/SM used to simulate solid particle-laden turbulent flows primarily differ in their treatment of the continuous carrier phase whereas both treat the dispersed phase in a similar fashion, *i.e.* by tracking a large number of particles in a Lagrangian frame. LES attempts to produce a real turbulence of the large scales, whereas in RANS/SM, the stochastic simulation of turbulence is mainly designed for the dispersed phase and provides a synthetic turbulence along the particle path.

Lagrangian LES of particle-laden turbulent flows is generally sensitive to many issues whose effects on the final results are often crucial. Some of these issues are intimately linked to LES such as the accuracy of the subgrid scale SGS model used to model the effect of the small scales on the large scales of motion. Other issues are generic to the simulation of particle-laden turbulent flows. For instance, the accuracy of the interpolation schemes used to compute the fluid velocity at the particle position and the one with which terms representing forces acting on particles in the Lagrangian particle equation of motion are written.

To compute the particle trajectory using one of the equations of particle motion discussed in section (3.4), the complete information of the carrier-phase instantaneous velocity is needed. The LES velocity field computed from the closed filtered Navier-Stokes equation contains part of the information. The remaining information is contained in the subgrid scale velocity field that is discarded by the filtering operation. The LES computation provides only the statistical properties of this small-scale velocity field. Consequently, with LES the information regarding the small scales is lost. This could be a severe loss for particles with smaller relaxation times compared to SGS turbulence timescales, *i.e.*, for particles more responsive to subgrid scale motions occurring on smaller timescales. Thus, a more complete approach should be the one with the inclusion of SGS motion effects on particle transport.

In the next Chapter, a Langevin-type stochastic diffusion process is adopted and developed to model the Lagrangian instantaneous fluid velocity seen by the solid particles u_s along their trajectories for hybrid Lagrangian-Eulerian LES of particle-laden turbulent flows. The model is based on the Eulerian instantaneous filtered velocity field and its statistics are provided by the LES of the continuous phase. The model accounts for both particle inertia and cross trajectory effects due the presence of a body force. These two effects cause the fluid element and particle trajectories to differ. Particle inertia induces a relative instantaneous motion of particles with regard to their fluid neighborhood. This is due to the modified time scale with which an inertial particle sees the turbulence around it. This time scale of the fluid velocities seen by the inertial particles can be regarded as changing between two asymptotic limits as function of particle response time τ_p. Indeed inertial particle with negligible inertia tend to follow their surrounding fluid elements and therefore see velocities which are correlated in a time interval of the order of the Lagrangian time scale. On the contrary, inertial particles with very high inertia are nearly at standstill with respect to motion of fluid elements and tend to see fluid velocities at about the same point whose integral time scale is the Eulerian one. Cross trajectory effects are the cause of external forces that induce a mean drift between the discrete particles and the surrounding fluid. Therefore, a separation of the average trajectories of the discrete and of the fluid elements occurs. The result of that is a de-correlation of the velocity of the fluid seen with respect to the velocity of fluid particles. Therefore the time scale with which inertial particles see the turbulence decreases.

It is worthwhile remembering that it is possible to decompose this driving fluid velocity into the filtered or resolved part and the SGS fluctuating part. The resolved fluid velocity field results from the continuous phase computation using LES and is interpolated at the particles locations. Then, the Lagrangian stochastic model gives the SGS fluctuating component. This decomposition is not necessarily the most convenient one [29]. Generating the instantaneous velocities, as it is the case for this study, is often better in the general case. This is mainly to avoid the spurious drift. Indeed, when Langevin equation is used to generate the fluctuating component, the fluctuating pressure gradient is taken into account through terms that are not closed [3]. Modeling these terms could yield a spurious drift. In the case of the instantaneous component, the mean pressure field is provided by the Eulerian simulation. The result is an instantaneous seen velocity field free of any spurious drift that takes into account the absent subfilter motion [30].

DISCLOSURES

Part of information included in this chapter has been previously published in *International Journal of Heat and Fluid Flow* 2008, Volume 29, Issue 4, pp. 1010-1028.

REFERENCES

[1]   Clift R, Grace JR, Weber ME. Bubbles, drops and particles. London Academic Press 1978; 380 pp.

[2]   Mashayek F, Pandya RVR. Analytical description of particle/droplet-laden turbulent flows. Prog Energy Combust Sci 2003; 29(4); 329-378.

[3]   Minier JP, Peirano E. The PDF approach to turbulent polydispersed two-phase flows. Phys Rep 2001; 352; 1-214.

[4]   Berlemont A, Desjonqueres P, Gousebet G. Particle Lagrangian simulation in turbulent flows. Int J Multiphase Flow 1990; 16(1); 19-34.

[5]   Crowe CT, Trout TR, Chung JN. Numerical models for two-phase turbulent flows. Annu Rev Fluid Mech 1996; 28; 11-43.

[6]   Gousebet G, Berlemont A. Eulerian and Lagrangian approaches for predicting the behavior of discrete particles in turbulent flows. Prog Energy Combust Sci 1999; 25; 113-159.

[7]   Lakehal D. On the modeling of multiphase turbulent flows for environmental and hydrodynamic applications. Int J Multiphase Flow 2002; 28; 823-863.

[8]   Loth E. Numerical approaches for motion of dispersed particles, droplets and bubbles. Prog Energy Combust Sci 2000; 26; 161-223.

[9]   Maxey MR, Riley JJ. Equation of motion for a small rigid sphere in a nonuniform flow. Phys Fluids 1983; 26(4); 883-9.

[10]   Crowe CT, Sommerfeld M, Tsuji Y. Multiphase flows with droplets and particles. CRC Press Boca Raton FL. 1998.

[11]   Williams MMR, Loyalka SK. Aerosol science: theory and practice. Pergamon Press Oxford 1991.

[12]   Crowe CT. On models for turbulence modulation in fluid-particle flows. Int J Multiphase Flow 2000; 26; 719-727.

[13]   Gore RA, Crowe CT. The effect of particle size on modulating turbulent intensity. Int J Multiphase Flow 1989; 15; 279-285.

[14]   Hosokawa S, Tomiyama A. Turbulence modification in gas-liquid and solid-liquid dispersed two-phase pipe flows. Int J Heat Fluid Flow 2004; 25; 489-498.

[15]   Hadinoto K, Jones EN, Yurteri C, Curtis JS. Reynolds number dependence of gas-phase turbulence in gas-particle flows. Int J Multiphase Flow 2005; 31; 416-434.

[16]   Maeda M, Hishida K, Furutani T. Optical measurement of local gas and particles velocity in an upward flowing dilute gas-solid suspension. Poly-phase Flow Transport Technology, Century 2-ETC, San Francisco, CA 1980.

[17]   Eaton JK, Segura JC. On momentum coupling methods for calculation of turbulence attenuation in dilute particle-laden gas flows. Report TSD-135 2005.

[18]   Squires KD, Eaton JK. Particle response and turbulence modification in isotropic turbulence. Phys Fluids 1990; A(2); 1191-1203.

[19]   Lomholt S, Stenum B, Maxey MR. Experimental verification of the force coupling method for particulate flow Int J Multiphase Flow 2002; 28; 225-246.

[20]   Sundaram S, Collins LR. A numerical study of the modulation of turbulence by particles. J Fluid Mech. 1999; 379; 105-143.

[21]   Boivin M, Simonin O, Squires KD. Direct numerical simulation of turbulence modulation by particles in isotropic turbulence. J Fluid Mech 1998; 375; 235-263.

[22]   Segura JC. Ph.D. thesis. ME Dept. Stanford University. California 2004.

[23]   Paris T, Eaton JK. Turbulence attenuation in a particle-laden channel flow. Report TSD-137. CTR, Stanford University 2001.

[24]   Elghobashi SE. On predicting particle-laden turbulent flows. Appl Sci Res 1994; 52; 309-29.

[25]   Tchen CM. Mean value and correlation problems connected with the motion of small particles suspended in a turbulent fluid. PhD Thesis, Delft University, The Hague 1947.

[26]   Saffman PG. The lift on a small sphere in a slow shear flow. J Fluid Mech 1968; 31; 624.

[27]   Stone HA. Philip Saffman and viscous flow theory. J Fluid Mech 2000; 409; 165-183.

[28]   Arnason G, Stock DE. Dispersion of particles in turbulent pipe flow. American Society of Mechanical Engineers. Fluid Engineering Division 1983; 10; 25-29.

[29]   Pozorski J, Minier JP. On the Lagrangian turbulent dispersion models based on the Langevin equation Int J Multiphase Flow 1998; 24; 913-945.

[30]   Mc Innes JM, Bracco FV. Stochastic particle dispersion modeling and the tracer-particle limit. Phys Fluids 1992; A(4); 2809.

CHAPTER 4

Lagrangian Stochastic Modeling of Subfilter Motion for LES of Particle-Laden Flow

Abstract: In this chapter, the adoption and development of a Langevin-type Lagrangian stochastic model for the transport of inertial particles by subfilter motion in LES is detailed. The different assumptions and numerical requirements needed for the implementation of the model are discussed.

The theoretical and numerical formulations of the Langevin model have been extensively discussed in the framework of particle-laden RANS. Its use is extended herein with the necessary modifications for the modeling of the fluid velocity seen by particles in LES framework. We introduce below the two formulations of Langevin equation used to model the time increment of the fluid velocity seen by inertial particles. For the first formulation, called *standard formulation*, closure of both drift and diffusion terms are similar to the one used for the fluid particle case. However, the SGS time scale with which inertial particles see the turbulence is modified to account for inertia and cross trajectories effects. For the second formulation, *referred to as complete formulation*, closure forms for the drift and the diffusion terms are described.

Numerical issues linked to the solution of the resulting stochastic differential equations as well as some important properties of the Langevin-type equations are discussed.

Keywords: Large eddy simulation; multiphase flows; turbulent flows; Lagrangian modeling; stochastic modeling, Langevin equation, drift term; diffusion matrix; spurious drift; fluid velocity seen.

4.1 INTRODUCTION

In this chapter, the adoption and development of a Langevin-type Lagrangian stochastic model for the transport of inertial particles by subfilter motion in LES is detailed. The different assumptions and numerical requirements needed for the implementation of the model are discussed. We describe first the use of the Langevin equation for single-phase turbulent flows.

The Lagrangian stochastic approach has been widely used for the simulation of single-phase turbulent flows over the past two decades. In this approach, a large number of fluid particles that have their own set of properties (position, velocity, composition, etc) are used to represent the turbulent flow. The time evolution of these fluid particle properties is based on stochastic model equations that could simulate the fluid particles. The archetypical stochastic model is developed according to the Langevin equation which is statistically equivalent to the Fokker-Plank equation governing the Probability Density Function (PDF). In the generalized Langevin model used to simulate the motion of the fluid particle, the position and velocity of the fluid particle evolve by [1]:

$$dx_{f,i} = u_{f,i}.dt,$$

$$du_{f,i} = (-\frac{1}{\rho f}\frac{\partial <p>}{\partial x_i} + \frac{1}{Re}\frac{\partial^2 \bar{u}_{f,i}}{\partial x_j \partial x_j} - \frac{u_{f,i}-<u_{f,i}>}{T_L}.dt + \sqrt{C_0 <\epsilon>}\,dW_i. \tag{4.1}$$

In these equations, T_L is the Lagrangian time scale, $<\epsilon>$ the dissipation rate of the kinetic energy k, C_0 is a model constant and W_i denotes Weiner process [2].

The PDF approach can be of practical use for a simulation approach only when a Monte-Carlo scheme [3] is implemented to solve the statistically equivalent stochastic differential equations. The PDF approach, which has originally been implemented in RANS, is generalized for the LES applications through the Filtered Density Function (FDF) approach [4, 5].

For the simulation of two-phase flows, modeling the evolution of the fluid velocity seen by the inertial particle along their paths can be achieved using a Langevin-type equation similar to that implemented in the Lagrangian Monte-Carlo

Abdallah Sofiane Berrouk

simulation of single phase flows [1]. In this approach, the solid particle position and velocity and the velocity of the fluid particle seen compose a random vector which evolves based on a Langevin-type equation. The obtained Langevin equation is statistically equivalent to the PDF Fokker-Plank equation where the components of the random vector are the internal variables $P = P(x_{p,i}, u_{p,i}, x_{s,i})$. It is worthy to note that the statistics of the fluid particle seen differs from those of the fluid particle due to the solid particle inertia and the crossing trajectory effects that need to be included.

4.2. Modeling of Fluid Velocity Seen by Inertial Particles

Langevin models [6] have been attractive stochastic diffusion models developed for fluid particle turbulent velocities [7]. They have been extended for the generation of the fluid turbulent field seen by inertial particles. The general form of the Langevin model chosen for the velocity of the fluid seen by particles is:

$$du_{s,i} = A_{s,i}(t, x_p, u_p, u_s)dt + B_{s,ij}(t, x_p, u_p, u_s)dW_j,$$
(4.2)

where the drift vector A and the diffusion matrix B have to be modeled. Each component of the vector dW is a Wiener process (white noise); it is a stochastic process of zero mean, $< dW >= 0$, a variance equal to the time interval, $< (dW)^2 >= dt$, and delta-correlated in the time domain [2].

The theoretical and numerical formulations of the Langevin model have been extensively discussed in the framework of particle-laden RANS [8, 9, 10]. Its use is extended herein with the necessary modifications for the modeling of the fluid velocity seen by particles in LES framework. We introduce below the two formulations of Langevin equation used to model the time increment of the fluid velocity seen by inertial particles. For the first formulation, called *standard formulation*, closure of both drift and diffusion terms are similar to the one used for the fluid particle case. However, the SGS time scale with which inertial particles see the turbulence is modified to account for inertia and cross trajectories effects. For the second formulation, *referred to as complete formulation*, closure forms for the drift and the diffusion terms are described.

4.2.1. Standard Formulation

The standard formulation of the time increment of the fluid velocity seen by inertial particles is expressed as follows [8]:

$$dx_{p,i} = u_{p,i}dt,$$

$$du_{p,i} = \frac{u_{s,i} - u_{p,i}}{\tau_p}.dt + g_idt,$$

$$du_{s,i} = \left(-\frac{1}{\rho f}\frac{\partial \bar{p}}{\partial x_i} + \frac{1}{Re}\frac{\partial^2 \bar{u}_{f,i}}{\partial x_j \partial x_j}\right).dt - \left(\frac{u_{s,i} - \bar{u}_{f,i}}{T^*_{SGS,i}}\right).dt + \sqrt{C_0 < \epsilon_r >} \, dW_i.$$
(4.3)

Here, ϵ_r is the dissipation rate of the residual or subfilter turbulent kinetic energy k_r, and C_0 is the Kolmogorov constant. $T^*_{SGS,i}$ is time scale with which inertial particles see the SGS turbulence. g_i is the gravity vector.

For LES, the Lagrangian time scale for the subgrid fluctuations $T_{L,SGS}$ is computed using the subfilter kinetic energy k_r and its dissipation rate ϵ_r. It reads [11]:

$$T_{L,SGS} = (\frac{1}{2} + \frac{3}{4}C_0)^{-1}\frac{k_r}{<\epsilon_r>}.$$
(4.4)

In the case of the Smagorinsky model, if equilibrium is assumed at the cutoff, the production balances the dissipation. Thus, the dissipation rate and the residual kinetic energy can be evaluated as the following:

$$\epsilon_r = -\tau_{ij}.\frac{d\bar{u}_i}{dx_j} = (C_s.\Delta)^2|\bar{S}|^3,$$
(4.5)

$$k_r = C_\epsilon(\Delta.\epsilon^r)^{2/3}.$$
(4.6)

Typically, $C_\epsilon \approx 1$ and $C_0 \approx 2.1$,[4].

The directional dependence of the fluid Lagrangian SGS time scales $T_{L,SGS}$ is neglected since subgrid scales are assumed to be homogeneous and isotropic. The Carlier *et al.* [12] model that takes into account the anisotropic character of the turbulent fluctuations is frequently used in RANS as we shall see in Chapter 7.

The fluid Lagrangian time scale seen by inertial particles T^*_{SGS} is $T_{E,SGS}$ (the Eulerian SGS time scale) in the limit of large Stokes number. On the other hand, $T^*_{SGS} = T_{L,SGS}$ (the Lagrangian SGS time scale) if $St \to 0$ since in this case the inertial particles reduce to fluid elements. Thus, in general T^*_{SGS} is a function of St and varies between $T_{L,SGS}$ and $T_{E,SGS}$ as it is portrayed in the following equation [13]:

$$T^*_{SGS} = \frac{T_{L,SGS}}{\beta}\left(1 - (1 - \beta)(1 + St)^{-0.4(1+0.01St)}\right), \tag{4.7}$$

St is Stokes number based on the Eulerian SGS time scale $T_{E,SGS}$ and β is the ratio between the Lagrangian and the Eulerian time scales. It is assumed that β keeps the same value across the different scales of turbulence:

$$\beta = T_L/T_E = T_{L,SGS}/T_{E,SGS}. \tag{4.8}$$

It was shown that the value of β is Reynolds number dependent [14] and varies considerably in the literature. When formula (4.7) is used to take into account the inertia effect, β is chosen to be 0.356 [13]. Two other values are tested: $\beta = 1.3$ [15] and $\beta = 0.8$ [14].

Equation (4.7) was developed for particles interacting with homogeneous and isotropic turbulence. Its use to account for inertia effect on Lagrangian subgrid time scale is more appropriate compared to its use to include inertia effect on Lagrangian time scale T_L in the framework of RANS/SM. For the latter approach, the construction of a wide spectrum of anisotropic turbulence fluctuations is sought through the stochastic modeling.

To account for the cross trajectory effect due to the presence of a body force, the Lagrangian SGS time scale is expressed in the case of inertial particle as function of the mean relative velocity between the fluid and the inertial particles. Since there is a difference between the longitudinal and transverse length scales for the spatial correlations, it is important to have a non isotropic form for T^*_{SGS}. To achieve this purpose, it is important to distinguish between the time scale measured in the direction of the mean relative velocity and the time scale measured in the transversal direction. If the direction (1) is aligned with the direction of the mean relative velocity and (2) and (3) are the transversal ones, Csanady formulae [16] can be used to compute the different anisotropic time scales:

$$T^*_{SGS,\parallel} = \frac{T^*_{SGS}}{\sqrt{1 + \beta^2 |< u_r >|^2/(2k/3)}},$$

$$T^*_{SGS,\perp} = \frac{T^*_{SGS}}{\sqrt{1 + 4\beta^2 |<u_r>|^2/(2k/3)}} \tag{4.9}$$

Here $< u_r >$ is the mean slip velocity between fluid and inertial particles. In fact Csanady formulae also take into account the continuity effect. The continuity effect postulates that the inertial particle dispersion in a direction perpendicular to the mean drift is twice as faster as inertial particle dispersion in a direction parallel to the mean drift.

The model depicted by the stochastic differential system (4.3) is in general sufficient to compute the "seen fluid velocity" when the interaction between the inertial particle and turbulence is weak, that is when the inertial effects are more important than the turbulent effects. It can be sufficient also if the injection conditions of the solid particles and their interactions with boundary conditions (wall collision) dominate the particle displacements. However, this stochastic model suffers many theoretical drawbacks [10]. The time scale $T^*_{SGS,i}$ is not considered in this model as macroscopic characteristics since it depends on the instantaneous velocities. The macroscopic effect usually reduces the effective time scale [17]. Moreover, this model is very simple and considers independent particles and consequently it does not carry any information concerning the mean behavior of a cloud of particles.

4.2.2. Complete Formulation

To cope with the drawbacks of the standard formulation, a complete formulation of Langevin-type model is used. In this model the drift coefficient and the diffusion matrix have more complicated expressions. For the drift terms, the mean statistical behavior of the cloud of particles that are present in one cell at time t is taken into account [9]. This results in an additional term in the equation of the drift. For the diffusion term, the diffusion constant C_0 (Kolmogorov constant) has a more complicated form taking into account the cross trajectory effects due to the presence of a body force and the extra viscous dissipation due to the subgrid scales.

Closure of the drift term

In the fluid case, the drift term entering the stochastic differential equation (4.2) is chosen to be a sum of a filtered term and a subgrid fluctuation term that is characterized by a time scale $T_{SGS,i}^*$. This form is retained for the two-phase flow case with a modification of both filtered and fluctuating terms to account for the inertia and cross trajectory effects. The filtered term is obtained by performing a first-order Taylor expansion for small dt and thus for small relative displacement $((< u_{p,j} > -\bar{u}_{f,j})dt)$. This term accounts for the average behavior of a cloud of particles present in one cell at time t. The fluctuating term is written as a return-to-equilibrium term with a modified subgrid time scale $T_{SGS,i}^*$.

$$A_s = \left(-\frac{1}{\rho_f}\frac{\partial \bar{p}}{\partial x_i} + \frac{1}{Re}\frac{\partial^2 \bar{u}_i}{\partial x_j \partial x_j}\right) + \left(< u_{p,j} > -\bar{u}_{f,j}\right)\frac{\partial \bar{u}_{f,i}}{\partial x_j}.dt - \left(\frac{u_{s,i}-\bar{u}_{f,i}}{T_{SGS,i}^*}\right). \tag{4.10}$$

Here $\bar{p}$ and $\bar{u}$ are the filtered pressure and velocity fields respectively. $T_{SGS,i}^*$ is the fluid Lagrangian time scale of the subgrid fluctuations seen by inertial particles. Other forms can be used but the present one, although postulated, is consistent with the general description of the cross-trajectory effect.

Closure of the diffusion term

Closing the diffusion matrix is worked out in several steps, by considering first the case of stationary and isotropic turbulence. By direct application of stochastic calculus, the stationarity constraint implies that $B_s^2 = 2u_s^2/T^*$. Using the isotropic assumption, $k = \frac{3}{2}u_s^2$, and the expression of T^* in the stationary case, $T^* = 4k/(3C_0^* b_i < \epsilon_r >)$, an expression of the diffusion term is derived:

$$B_s = \sqrt{C_0^* b_i < \epsilon_r >}, \tag{4.11}$$

$$b_i = T_{SGS}^*/T_{SGS,i}^* \text{ is the correction factor in Csanady's formulae (Eqns. 4.9).}$$

For the case of stationary fluid velocity seen but not necessarily isotropic the above closure of diffusion term is not satisfactory. This is due to the fact that the drift term in this case is not isotropic. Thus contributions coming from drift and diffusion terms do not balance to ensure that the stationarity constraint concerning the kinetic energy of the fluid seen is respected, $du_s^2 = 0$. To resolve this difficulty, a modified residual or SGS kinetic energy is introduced:

$$\tilde{k}_r = \frac{3}{2}\frac{\sum_{i=1}^3 b_i <u_i'^2>}{\sum_{i=1}^3 b_i}. \tag{4.12}$$

It represents the normal energies weighted by the corresponding Csanady's factors b_i. Since these factors are anisotropic, the modified kinetic energy $\tilde{k}_r$ differs from the plain one k_r. For isotropic turbulence, there is no difference between $\tilde{k}_r$ and k_r. This is also the case when fluid particles are considered, or when no cross-trajectory effects come into play. Using the weighted residual or SGS kinetic energy of the fluid seen, the diffusion term is modified:

$$B_s = \sqrt{< \epsilon_r > (C_0 b_i \frac{\tilde{k}_r}{k_r} + \frac{2}{3}(b_i \frac{\tilde{k}_r}{k_r} - 1)}. \tag{4.13}$$

The next step is to consider the general case of a non-stationary turbulence. A new constraint has to be taken into account; $du_s^2 = -2 < \epsilon_r > dt$. The viscous dissipation term is accounted for by adding a negative term, $-2/3 <$

$\epsilon_r >$ in the diffusion term. The expression of the time scale T_{SGS}^* is also slightly modified. It is computed in the non-stationary case using Eqn. (4.4). Thus, the value of the proportionality constant C_0^* is now linked to the Kolmogorov constant C_0: $C_0^* = C_0 + 2/3$. This results in the following diffusion term:

$$B_s = \sqrt{< \epsilon_r > (C_0 b_i \frac{\tilde{k}_r}{k_r} + \frac{2}{3} (b_i \frac{\tilde{k}_r}{k_r} - 1)}. \qquad (4.14)$$

In the case of the complete formulation, the diffusion matrix B_s is derived by taking into account the mean drift of inertial particles caused by the body force and the extra turbulent energy dissipation induced by the subgrid scale motion. It is important to emphasize that the resulting diffusion matrix is diagonal but non-isotropic: $B_{s,ij} = B_{s,i} \delta ij$. This allows to retrieve the fluid particle case when $b_i = 1$.

The Langevin model for the time increments of the fluid velocity seen by inertial particles is obtained by adding the drift and the diffusion terms developed above:

$$dx_{p,i} = u_{p,i}.dt,$$

$$du_{p,i} = \frac{u_{s,i} - u_{p,i}}{\tau_p}.dt + g.dt, \qquad (4.15)$$

$$du_{s,i} = \left(-\frac{1}{\rho_f}\frac{\partial \bar{p}}{\partial x_i} + \frac{1}{Re}\frac{\partial^2 \bar{u}_i}{\partial x_j \partial x_j}\right)dt + \left(< u_{p,j} > -\bar{u}_{f,j}\right)\frac{\partial \bar{u}_{f,i}}{\partial x_j}dt - \left(\frac{(u_{s,i} - \bar{u}_{f,i})}{T_{SGS,i}^*}\right)dt.$$

$$+\sqrt{C^d < \epsilon_r >}\, dW_i.$$

The diffusion coefficient C^d is evaluated according to the following formulation:

$$C^d = C_0 b_i \tilde{k}_r / k_r + \frac{2}{3}\left(b_i \tilde{k}_r / k_r - 1\right). \qquad (4.16)$$

4.3. Two-Way Coupling Case

In case of two-way coupling, it is assumed that the presence of solid particles changes the transfer rate of energy and energy dissipation. However, the nature and structure of turbulence remain the same. Based on that assumption, Minier *et al.* [18] proposed a model for the seen fluid velocity in case of two-way coupling in RANS framework. An acceleration term is added to account for the presence of particles while the modeling of the drift vector and the diffusion matrix is kept the same since they are not affected by the nature of turbulence. However, Boivin *et al.* [19] indicated that the presence of particles in isotropic turbulence yield a non-uniform distortion of the energy spectrum. This means that the nature and the structure of the energy transfer mechanisms of turbulence are modified by the presence of particles. In this case, modeling of drift vector and diffusion matrix has to be altered to take into account such a modification in turbulence nature. This appears to be not feasible since knowledge of the structure of turbulence in presence of particles is lacking [9]. In the context of RANS, Minier *et al.* [18] proposed to add a source term to the mean part of the drift term to account for the two-way coupling situations. If this idea is retained for the LES calculations, the fluid velocity seen by inertial particle in the case of two-way coupling can be written as the following:

$$dx_{p,i} = u_{p,i}.dt, \; du_{p,i} = \frac{u_{s,i} - u_{p,i}}{\tau_p}.dt + g.dt,$$

$$du_{s,i} = \left(-\frac{1}{\rho_f}\frac{\partial \bar{p}}{\partial x_i} + \frac{1}{Re}\frac{\partial^2 \bar{u}_{f,i}}{\partial x_j \partial x_j}\right).dt + \left(< u_{p,j} > -\bar{u}_{f,j}\right)\frac{\partial \bar{u}_{f,i}}{\partial x_j}.dt$$

$$+\frac{\alpha_p \rho_p}{(1 - \alpha_p)\rho_f}\frac{u_{p,i} - \bar{u}_{f,i}}{\tau_p}.dt - \left(\frac{u_{s,i} - \bar{u}_{f,i}}{T_{SGS,i}^*}\right).dt$$

$$+\sqrt{C^d <\epsilon_r>}\, dW_i.$$ (4.17)

4.4. Numerical Issues

The stochastic differential equation (SDE) systems ((4.3), (4.15) or (4.17)) are often integrated using an appropriate weak second-order integration scheme [20]. This numerical scheme accounts for the nature of the problem characterized by the presence of different time scales. This can lead to stiff equations when the smallest time-scale is significantly less than the time-step of the simulation. This point is crucial for physical and engineering applications, where various limit cases can be present at the same time in different parts of the domain or at different times. Because the turbulence problem has a multi-scale character, three time scales are considered: The observation time scale, Δt, and two physical time scales, the particle relaxation time, τ_p, and the time scale of the fluid velocity seen, T_{SGS}^*. When these scales go to zero, a hierarchy of stochastic differential systems is obtained. Both formulations of Langevin equation presented above degenerate to a stochastic model for turbulent diffusion when $\tau_p \to 0$, that is, the inertial particles behave like fluid particles. The different asymptotic cases of the stochastic model are presented in Appendix A.

The weak second-order integration scheme consists of a prediction step and a correction step. The prediction step is a weak first-order integration scheme (Euler scheme). This numerical scheme is derived by freezing the coefficients on the integration intervals and by resorting to Ito's calculus; it can be shown that SDE systems (4.3) and (4.15) (and SDE system (4.17) when two-way coupling is considered) have an analytical solution [20]. The analytical solutions for particle position x_p, velocity u_p and seen fluid velocity u_s are given in Appendix B.

To enforce the mean-continuity constraint, a pressure correction is adopted in a similar way to the classical pressure correction step for the Eulerian fluid velocity field [18].

The pressure correction is added as a potential ϕ:

$$< u_{p,i} >_c = < u_{p,i} >_p - \frac{\partial \phi}{\partial x_i}.$$ (4.18)

The potential ϕ is computed by solving the Poisson equation:

$$\frac{\partial}{\partial x_i}\left(\alpha_p \rho_p \frac{\partial \phi}{\partial x_i}\right) = \frac{\partial}{\partial t}(\alpha_p \rho_p) + \frac{\partial}{\partial x_i}\left(\alpha_p \rho_p < u_{p,i} >_p\right).$$ (4.19)

Then the mean velocity correction term is applied to each particle velocity to enforce the particle continuity equation (density is constant):

$$\frac{\partial}{\partial t}(\alpha_p \rho_p) + \frac{\partial}{\partial x_i}\left(\alpha_p \rho_p < u_{p,i} >\right) = 0.$$ (4.20)

4.5. Properties of the Langevin Equations

The resulting Langevin equation is believed to represent the simplest model for the LES of two-phase flow. It is important to emphasize that the closure relations for the drift and diffusion terms are derived based on different assumptions and they reflect modeling choices. Nevertheless, the present Langevin model has a number of properties that are important for its physical and mathematical soundness. These properties are related to non-Gaussian behavior and spurious drift.

In fact, for the case of homogeneous turbulence, the use of Langevin equation to model either the time increments of the fluid velocity seen by inertial particles or particle Lagrangian velocity increments results in a Gaussian stochastic process. This is simply due to the fact that both drift and diffusion are constant. This is not the case when inhomogeneous turbulence is considered as we shall discuss below.

In fact there is an assumption of a Gaussian form built within a Langevin model portrayed in Eqn. (4.2). This is due to the increments of the Wiener process which is independent Gaussian variables. In the general case this does not imply anything about the stochastic process u_s. Deviation from Gaussian behavior is not assumed beforehand in the

Langevin equation. Physically, it is the consequence of the mixing at a given location of particles or trajectories which have different histories. Indeed, in the general case of inhomogeneous turbulence, the different filtered quantities entering the drift and the diffusion terms are space dependent. Therefore, it is a must to consider the joint process of at least the particle position and the fluid velocity seen (x_p, u_s) if the statistics of the inertial particle velocity u_p are supposed known and constant. In this regard, the Langevin equation is a nonlinear stochastic process with respect to the variables of the joint process. Consequently, the resulting probability density function of the velocity of the fluid seen that comes out of the present Langevin model is not Gaussian.

Concerning spurious drift, Langevin models used to compute the time increments of fluid velocity seen do not suffer this drawback as it was demonstrated by Mc lnnes and Bracco [21]. Spurious drift has been a serious shortcoming for most of the Lagrangian models based on the *Random-Walk* approach and it was resolved by Pope's work [22]. Minier and Pierano [9] demonstrated that the Langevin model has two features that are needed to avoid spurious drifts. First, it is used in a Lagrangian formulation and all the usual terms arising from convection in the Navier-Stokes equation are implicitly contained in the Lagrangian derivative. Second, the mean pressure gradient is properly included in the particle velocity time evolution equation.

DISCLOSURES

Part of information included in this chapter has been previously published in *Journal of Turbulence* 2007, Volume 8, No.50, pp. 1-20

REFERENCES

[1]   Pope S.B. Turbulent flows. Cambridge University Press 2000.
[2]   Kloeden PE, Platen E, Schurz H. Numerical solution of SDE through computer experiment. Springer-Verlag 1994.
[3]   Oran ES, Oh CK, Cybyk BZ. Direct simulation Monte Carlo: recent advances and applications. Annu Rev Fluid Mech 1998; 30; 403-41.
[4]   Gicquel LYM, Givi P, Jaberi FA, Pope SB. Velocity filtered density function for large eddy simulation of turbulent flows. Phys Fluids 2002; A(14)3; 1196-1213.
[5]   Sheikhi MRH, Drozda TG, Givi P, Pope SB. Velocity-scalar filtered density function for large eddy simulation of turbulent flows. Phys Fluids 2003; A(15)8; 2321-2337.
[6]   Langevin P. Comptes Rendus. Acad Sci Paris 1908; 146; 530-533.
[7]   Haworth. DC, Pope S. A generalized Langevin model for turbulent flows. Phys Fluids 1986; 29(2); 387-405
[8]   Minier JP. Probabilistic approach to turbulent two-phase flows modeling and simulation: Theoretical and numerical simulations. Monte Carlo Methods Appl 2000; 7(3); 295-310.
[9]   Minier JP, Peirano E. The PDF approach to turbulent polydispersed two-phase flows. Phys Rep 2001; 352; 1-214.
[10]   Pozorski J, Minier JP. On the Lagrangian turbulent dispersion models based on the Langevin equation. Int J Multiphase Flow 1998; 24; 913-945.
[11]   Heinz S. Statistical mechanics of turbulent flows. Springer-Verlag Berlin 2003.
[12]   Carlier J. Ph, Khalij M, Oesterle B. An improved model for anisotropic dispersion of inertial particles in turbulent shear flows. Aerosol Sci Tech 2005; 39; 196-205
[13]   Wang LP, Stock DE. Dispersion of heavy particles by turbulent motion. J Atmos Sci 1993; 50(13); 1897-1913.
[14]   Sato Y, Yamamoto K. Lagrangian measurement of fluid-particle motion in an isotropic turbulent field. J Fluid Mech 1987; 175; 183.
[15]   Riley JJ. PhD dissertation, The Johns Hopkins University, Baltimore. Maryland 1971.
[16]   Csanady GT. Turbulent diffusion of heavy particles in the atmosphere. J Atmos Sci 1963; 20, 201-208.
[17]   Douce A. Modelisation stochastique Lagrangienne d'ecoulements turbulents diphasiques polydisperses dans Code_Saturne. HI-81/04/03/A. Internal Report EDF R&*D* 2004
[18]   Minier JP, Peirano E, Chibbaro S. PDF model based on Langevin equation for polydispersed two-phase flows applied to a bluff-body gas-solid flow. Phys Fluids 2004; 16(7); 2419-2431.
[19]   Boivin M, Simonin O, Squires KD. Direct numerical simulation of turbulence modulation by particles in isotropic turbulence. J Fluid Mech 1998; 375; 235-263.
[20]   Minier JP, Peirano E, Chibarro S. Weak first and second order numerical schemes for stochastic differential equations appearing in Lagrangian two-phase flow modeling. Monte Carlo Methods Appl 2003; 9(2); 93-133.

[21]   Mc Innes JM, Bracco FV. Stochastic particle dispersion modeling and the tracer-particle limit. Phys Fluids 1992; A(4); 2809.

[22]   Pope S.B. Consistency conditions for random walk models of turbulent dispersions. Phys Fluids 1987; 30(8); 2374-2378

Large Eddy Simulation of Solid Particle Dispersion in a Downward Turbulent Pipe Flow

Abstract: Numerical simulations using two approaches, namely RANS and LES, were conducted to compute inertial particle dispersion from a source point in a turbulent gas flow in a pipe at a high Reynolds number. Numerical predictions were compared to the experimental observations of Arnason (1982) and Arnason and Stock (1984). Stochastic modeling of the turbulent fluctuations seen by inertial particles along their trajectories has been used. In the framework of RANS, the aim is to reconstruct the whole turbulent field whereas in the context of LES, only modeling of SGS fluctuating velocities is sought.

Particle dispersion statistics such as particle concentration, radial velocity and the dispersion coefficient were computed for solid particles that have different inertia and drift. The use of a Langevin-type stochastic approach to model the sub-filter fluctuations has proven crucial for results concerning the small-Stokes-number particles. The stochastic model used has been extensively used in the framework of RANS. Its simplistic extension to predict the sub-filter fluctuations for LES has given very satisfactory results.

Numerical predictions show that, for the same flow, inertial particles with larger diameter (and hence larger Stokes number) can disperse faster than smaller particles (with smaller Stokes numbers). It was proved theoretically that this can be the case if the inertia parameter controls the dispersion. These findings back up the experimental observations of Arnason and Stock.

Comparison of RANS and LES results have shown that the RANS approach is unable to predict particle dispersion statistics as accurately as the LES in particular for inertial particles characterized by a Stokes number smaller than 0.5. For particles with Stokes number higher than 0.5, both LES and RANS predictions compare reasonably well with the experimental results.

Keywords: Pipe flows, Large eddy simulation, particle-laden flow, turbulence, Lagrangian description, small inertial particles, stochastic process, stokes number, dispersion; particle concentration.

5.1 INTRODUCTION

In this chapter the experimental work of Arnason [1] and Arnason and Stock [2] is simulated using two methodologies: RANS and LES. The experiment consists of injecting different sizes of solid particles in a downward turbulent pipe flow (see Fig. **5.1**) with geometrical and physical parameters summarized in Table (**5.1**). The measurements of particle dispersion coefficients were made possible by measuring particle number density and radial velocity profiles using the laser Doppler anemometer technique. Results at different pipe sections downstream the injection point were obtained for three sizes of glass beads dispersing from a point source on the pipe centerline. The physical characteristics of the solid particles are summarized in Table (**5.2**). Details of the experimental set up and data processing are given in Arnason [1] and Arnason and Stock [2].

This case study can be considered as a first step to predict particle dispersion occurring in many industrial processes such as spray drying, pulverized coal combustion and electrostatic precipitation of dust. Since dilute suspension of very small particles are considered, this case is an appropriate validation test of the Lagrangian stochastic model described in Chapter 4 in the framework of one-way coupling.

The main findings of this experimental work are very interesting. It was found that the $57\mu m$ particles disperse faster than the $5\mu m$ and $37\mu m$ particles. This fact is portrayed in Fig. (**5.2**). In fact such a trend that appears counter intuitive has also been observed in several previous experimental works on particle dispersion in pipe flows, in particular those of Calabrese *et al.* [3] and Jones [4]. Many explanations for this phenomenon have been given in the literature. Crowe *et al.* [5] indicated that particles with intermediate Stokes number would be dispersed significantly faster than the flow itself due to the centrifugal effects created by the organized vortex structures. Tang *et al.* [6] performed vortex simulations for a spatially developing plane mixing layer and wake laden with solid particles.

Abdallah Sofiane Berrouk

They showed that particle dispersion levels can greatly exceed fluid-particle dispersion at intermediate Stokes number. Also, theoretical analyses made by Nir and Pismen [7] and Reeks [8] predicted the possibility of solid particle dispersing faster than fluid particles in homogeneous and isotropic turbulence.

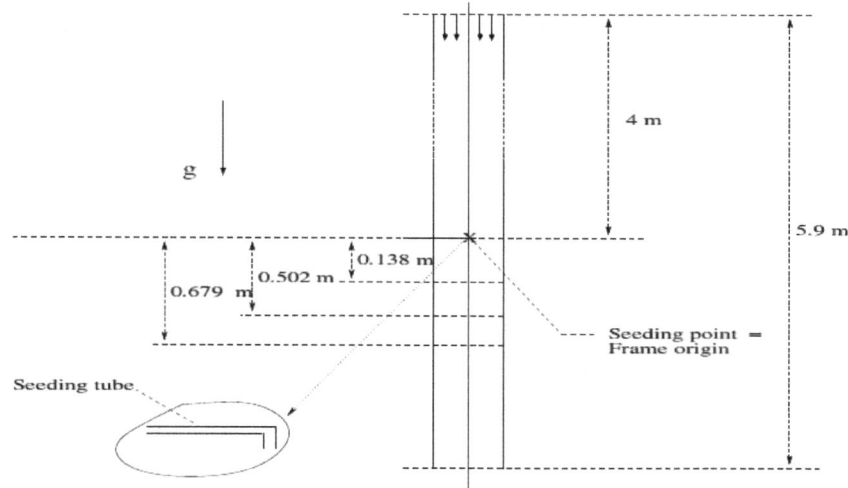

Figure 5.1: Schematic Description of the experimental layout.

Table 5.1: Data of Arnson's Experimental Work [6].

Pipe length	$L = 5.9\,m$
Pipe diameter	$D = 0.09\,m$
Maximal Velocity	$V_{max}\,9.56\,m/s$
Reynolds number based on maximal velocity	$Re = 50000$
Reynolds number based on friction velocity	$Re_\tau = 2200$

Table 5.2: Physical Characteristics of Inertial Particle Tracked in the Turbulent Pipe Flow.

Mean Diameter d_p^u (μm)	5	37	57
Standard deviation σ_p (μm)	1	8	11
Clipping (μm)	$2 < d_p^u < 10$	$18 < d_p^u < 75$	$32 < d_p^u < 101$
Density (kg/m^3)	2475	2450	2420
Response Time (ms)	0.2	10	25
Settling velocity (mm/s)	1.96	98	245

A mathematically simple and physically comprehensive analysis due to Wang and Stock [9] related the particle dispersion statistics to measurable flow statistics and particle parameters through simple algebraic equations. These algebraic equations were derived for the normalized particle diffusivity, rms fluctuating velocity, and Lagrangian time scale. It was concluded that in homogeneous isotropic flows, a solid particle disperses faster than a fluid element if the inertia controls the dispersion. They disperse slower than the fluid elements if the dispersion is governed by the drift due to the presence of an external force.

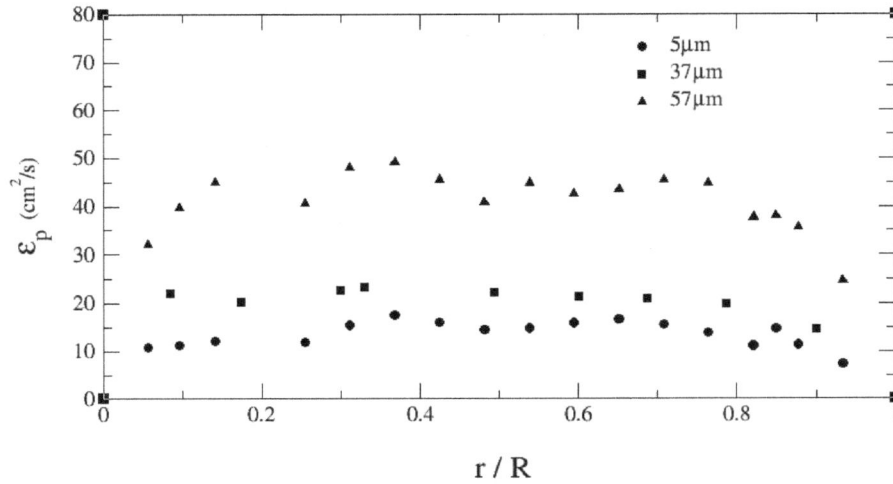

Figure 5.2: Experimental results of particle dispersion coefficients ϵ_p against the non-dimensionalised pipe radius (r/R) for the 5μm, 37μm, and 57μm particles.

An explanation of the findings of Arnason and Stock's experiment [2] can be given based on Taylor's theory [10] for particle dispersion taking into account both inertia and cross trajectory effects associated with a solid particle under the influence of a body force. In Taylor's approach, particle dispersion in isotropic, homogeneous and stationary turbulence is linked to the particle rms fluctuating velocity v_p' and the particle Lagrangian time scale T_{Lp} :

$$\epsilon_p = v_p'^2 . T_{Lp}. \tag{5.1}$$

Particle inertia causes the particles to respond sluggishly to fluid turbulence with particles not following turbulent fluctuations with small time scales. Increased particle inertia will reduce v_p' and increase T_{Lp} and thus ϵ_p can increase or decrease with increasing inertia depending on the relative variation of v_p' and T_{Lp}. The crossing trajectory effect (CT) takes place when an external force affects inertial particles in a different manner than the adjacent fluid particles due to a difference in density or electric charge. The external force will cause particle paths to deviate from the paths of surrounding fluid element. This causes particle to diffuse less than fluid elements. The CT effect will force particles to leave eddies prematurely and reduce T_{Lp}. This results in a reduced particle diffusivity.

From this simplistic description of particle dispersion we can conclude that crossing trajectory effects results always in a reduction of particle diffusivity while inertia effects can yield to an increase or decrease in particle diffusivity depending on the relative variation in both v_p' and T_{Lp} due to a variation in inertia. Let's consider two solid particles $P1$ and $P2$ having the same density but different diameters $d_{p1} < d_{p2}$ which yields $\tau_{p1} < \tau_{p2}$. Let's assume that both particles respond to at least a portion of the turbulent fluctuations present in the flow with different time scales, that is $\tau_{p1} < \tau_{p2} < T$, with T is the largest time scale of turbulent fluctuations that are seen by the two particles. In the absence of a body force and as a results of inertia, $v_{p1}' > v_{p2}'$ since $P1$ responds to much more turbulent fluctuations. However $T_{Lp2} > T_{Lp1}$ since $P2$ keeps longer the turbulent fluctuations transmitted to the particle by the turbulent fluid motion. As consequence no conclusion can be made from (5.1) on which particle would disperse faster. In presence of a body force, the crossing trajectory effects comes into play yielding a decrease of the time scale seen by solid particles (T becomes T^* with $T^* < T$). In case when $\tau_{p1} < T^* < \tau_{p2}$, particle $P2$ will not be influenced by any turbulent fluctuations because its particle response time is larger than any seen turbulent time scale present in the flow, $v_{p2}' \to 0$. In this case, $P1$ certainly disperses faster than $P2$. If $T^* < \tau_{p1} < \tau_{p2}$, both particles $P1$ and $P2$ will travel in the flow like bullets and it is meaningless to talk about particle dispersion by turbulence.

To be able to quantify this qualitative analysis, it is important to link both inertia and cross trajectory effects to the fluid turbulent fluctuations and their time scales as it was done in Wang and Stock's analysis [9]. Two parameters can be derived. The inertia parameter or the Stokes number that quantifies the inertia effect ($St = \tau_p/T^*$) and the drift parameter that is the ratio of the drift velocity to the fluid rms fluctuating velocity ($Dr = \tau_p.q/u_f'$, where q is a body force per unit of mass and u'_f is the fluid rms fluctuating velocity). The drift parameter measures the CT effect. As a result of Wang and Stock's analysis [9], another parameter appeared in the algebraic equations. This parameter is called the turbulence structure. It is defined as the ratio of the eddy decay time to the turnover time ($m = T_E.u'_f/L_f$, where L_f is the fluid integral length scale). For homogeneous and isotropic turbulence, it is equal to 1.

Wang and Stock's quantitative analysis [9] was in total agreement with the qualitative analysis based on Taylor's theory [10]. Indeed, it was concluded from this analysis that at very large drift parameter, solid particle diffusivity is smaller than the fluid element diffusivity and insensitive to the inertia parameter. Particle Lagrangian time scale increases with increasing inertia while particle rms fluctuating velocity decreases. An increase in particle dispersion with an increase in inertia was found for small drift parameters (less than 1 for dispersion in a perpendicular direction to the drift and less than 2 for dispersion in a direction parallel to the drift). Wang and Stock [9] concluded that inertial particles are found to disperse faster than fluid elements if the inertia parameter controls the dispersion and slower than fluid elements if the drift parameter governs the dispersion.

This is the picture about dispersion of particles travelling in homogeneous isotropic flows. It can be used to explain the experimental results on particle diffusivity in pipe flow at least in the core region of the pipe far from the walls. In these regions, the flow can be considered quasi homogeneous and isotropic.

Yet, in the presence of the shear, the analysis of the turbulent dispersion of solid particles becomes more complicated since its effect on particle dispersion is not well understood. However, some intuitive analyses were made to explain the trend shown in some experimental works for particle dispersion in pipe flows. Hinze [11] has linked these results to two mechanisms. The first one is the overshoot effect due to the in-homogeneity of the mean flow field that may reduce the diffusivity of particles with small inertia. The second is called the filtering effect that occurs when the particle size become greater than the size of the smallest eddies. We shall demonstrate later that this is not the case of the experiment in hand and the largest particle tracked has a diameter much smaller than Kolmogorov length scale. Reeks [12] expected an increased diffusivity when the particle Reynolds number based on particle diameter and turbulence intensity $Re_{p,i} = d_p.u'/\nu$ is beyond the Stokes regime ($Re_{p,i} > 1$).

Fig. (5.3) shows RANS and LES predictions of particle Reynolds number based on particle diameter and turbulence intensity in the radial direction (radial rms fluctuating velocity). Both numerical simulations predicted a value around unity for $57\mu m$ particles. This can be responsible for their increased diffusivity compared to $37\mu m$ and $5\mu m$ particles. Lee and Durst [13] suggested that larger particles do not respond to the fluid turbulence, but are driven towards the walls by interaction with the mean velocity profile through the Saffman lift force. Arnason and Stock [2] made a rough calculation of the Saffman lift force for the $57\mu m$ particles. They indicated that it could not be responsible for the high migration rate. The turbulence structures can be also responsible of a reduced diffusivity of very small solid particles that are transported by almost all the eddies present in the pipe flow. Indeed, turbulent flows contains fluid eddies that have random velocities relative to the mean flow. To satisfy continuity, these eddies keep moving in opposite directions to each other giving rise to what can be called a trade area of eddies. It is highly likely that small solid particles (small inertia) can be trapped within this trade area of eddies impeding them to disperse towards the pipe wall. Larger solid particles have less chance to be brought back to the pipe center because of their increased inertia. Thus, they may disperse faster though they sense smaller portion of the turbulent fluctuations.

For want of a better understanding, some numerical predictions are anticipated to clarify this tendency in an instantaneous sense. Figs. (5.4) to (5.7) show instantaneous particle trajectories of some identical particles of each size class. They were followed from the injection point until they exit the pipe. Also, fluid elements trajectories are evaluated. For the sake of clarity, particles with a small angle of dispersion are not represented. These particles tend to disperse more slowly and stay in the pipe center or cross towards the other side of the pipe. They constitute a high percentage of injected particles and including their trajectories will influence the figures' readability. Trajectories are computed using the flow field generated by RANS and LES. Dotted lines represent trajectories of particles that have been dispersed by the turbulent motion without being brought back to the pipe center before hitting the walls.

Solid lines are trajectories of particles that have been trapped in the mentioned trade areas at some stage of their radial dispersion and consequently have seen their trajectories constantly changing directions which results in a decreasing dispersion. For both RANS and LES predictions, it is easy to notice that the smaller inertia the particle has, the more its radial dispersion is decreased due to a constant change in particle radial direction by the turbulent motion.

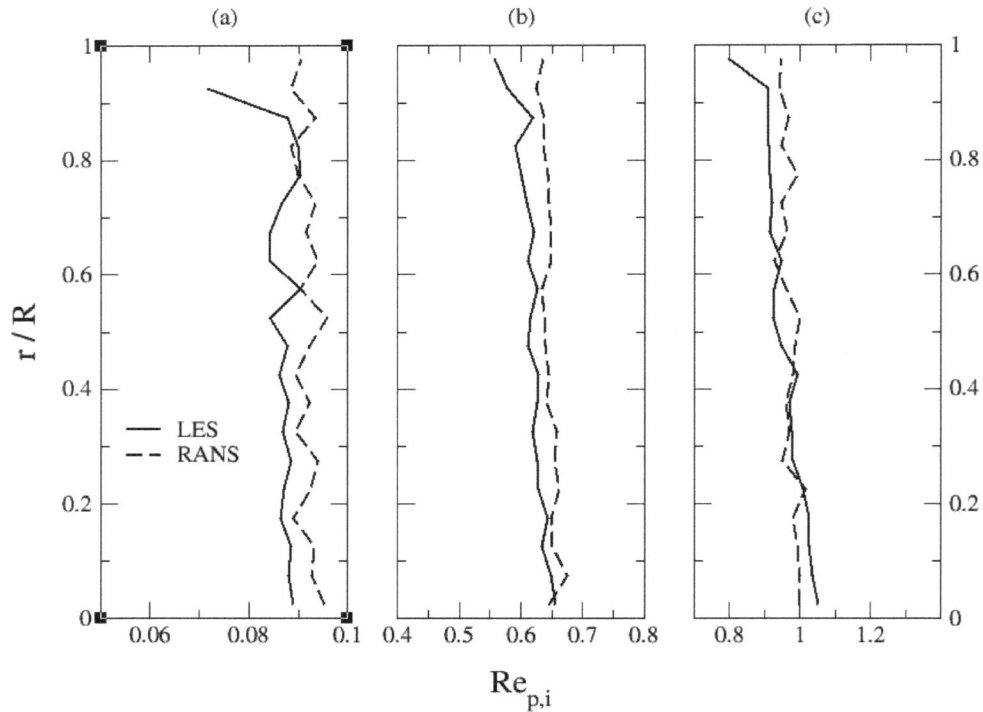

Figure 5.3: RANS and LE predictions of the particle Reynolds number based on turbulence intensity $Re_{p,i}$ versus the non-dimensionalised pipe radius (r/R). (a) $5\mu m$ particles, (b) $37\mu m$ particles, (c) $57\mu m$ particles.

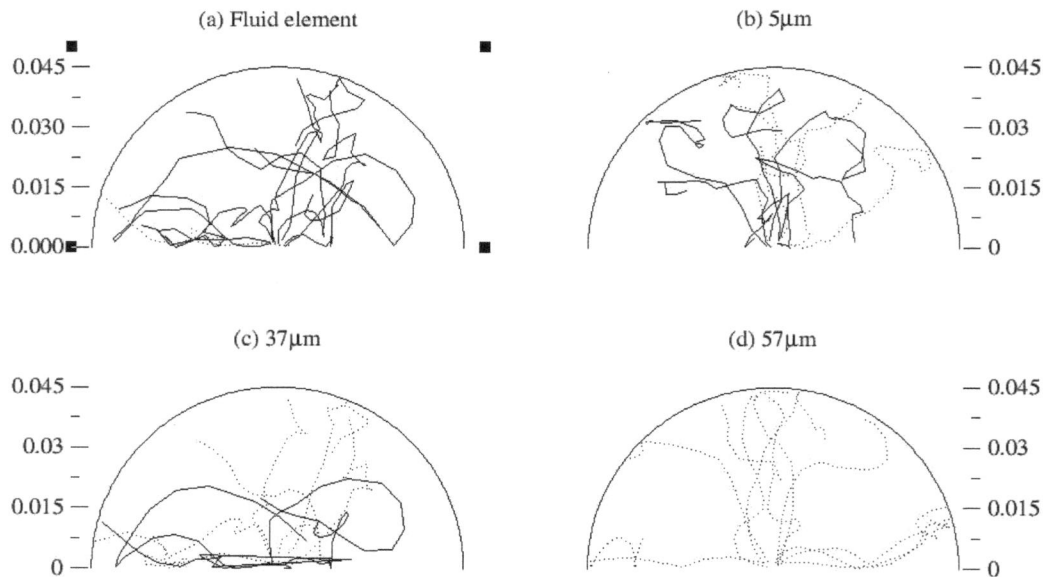

Figure 5.4: RANS predictions of the instantaneous particle trajectories. View in the pipe cross section. Pipe radius $R = 0.045m$. (a) Fluid particles, (b) $5\mu m$ solid particles, (c) $37\mu m$ solid particles, (d) $57\mu m$ soid particles.

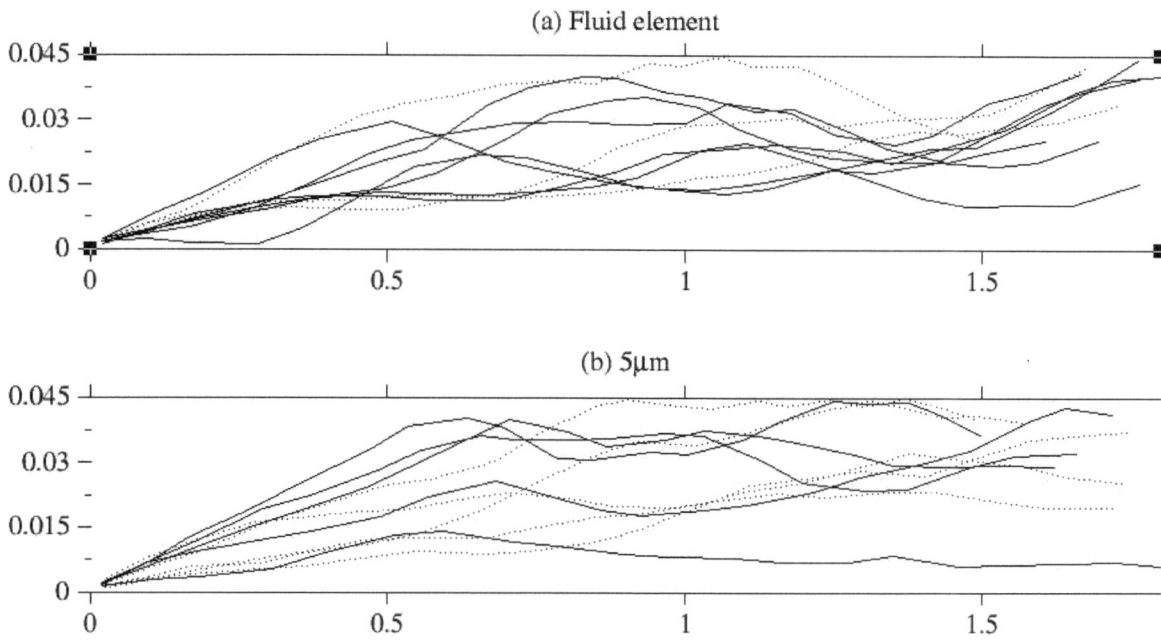

Figure 5.5: RANS predictions of the instantaneous particle trajectories. View in the streamwise direction: Pipe radius $R = 0.045m$, pipe length $L = 1.9m$. (a) Fluid particles, (b) $5\mu m$ particles, (c) $37\mu m$ particles, (d) $57\mu m$ particles.

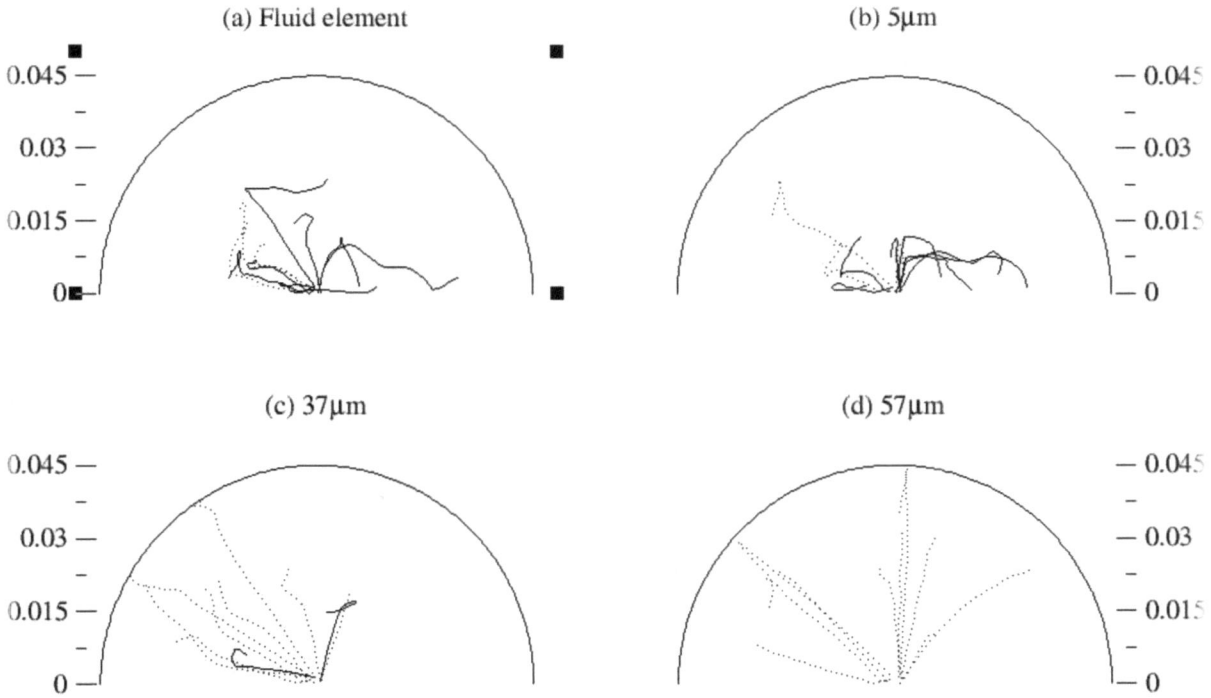

Figure 5.6: LES predictions of the instantaneous particle trajectories. View in the pipe cross section: Pipe radius $R = 0.045m$. (a) Fluid particles, (b) $5\mu m$ particles, (c) $37\mu m$ particles, (d) $57\mu m$ particles.

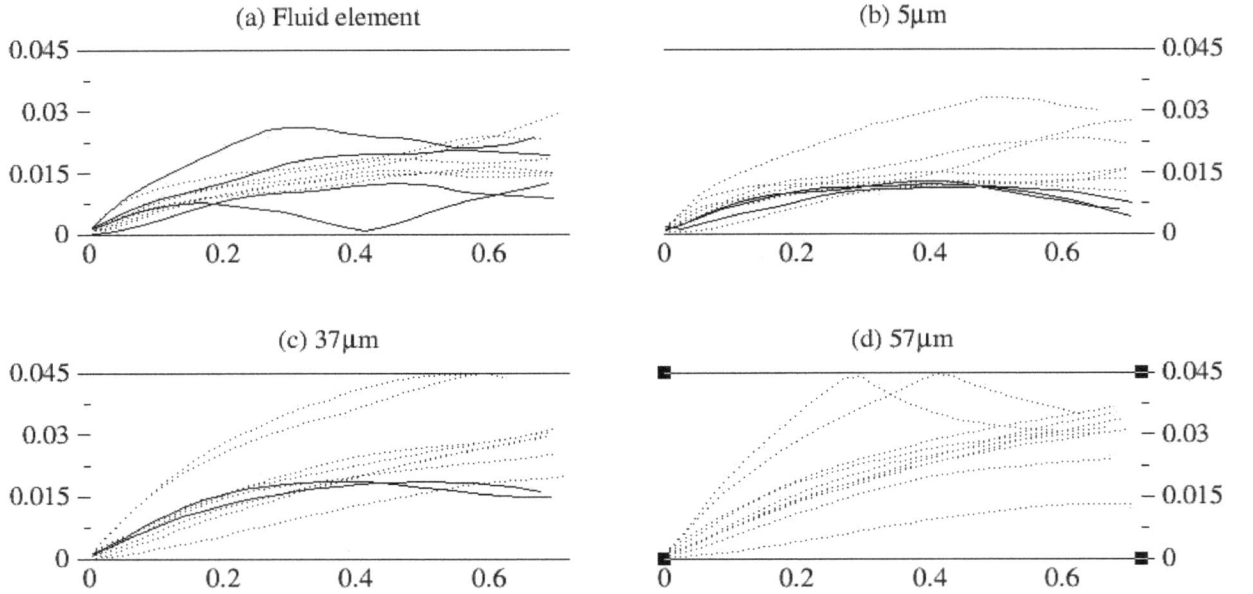

Figure 5.7: LES predictions of the instantaneous particle trajectories. View in the streamwise directions: Pipe radius $R = 0.045m$, pipe length $L = 0.72m$. (a) Fluid particles, (b) $5\mu m$ particles, (c) $37\mu m$ particles, (d) $57\mu m$ particles.

This tendency is also confirmed by the particle statistics as we shall discuss it later in this chapter. We can conclude that the seemed-contradictory Arnason and Stock's experimental results can be well explained by theoretical analysis developed for particle dispersion in homogeneous and isotropic turbulence and also by some physical considerations. Numerical simulations of the experiment can add more credibility to these experimental results and shed more light on the theoretical analysis by providing more information about both the carrier and dispersed phases.

5.2 THEORETICAL FORMULATION

5.2.1 Gas Phase

In order to simulate the experiment of Arnason [1], a turbulent pipe flow is studied in a Cartesian framework. The diameter of the pipe is denoted by D, and the length of the computational domain by L which is taken equal to $8D$ to include the largest-scale structure. This can be checked by ensuring that the fluctuating velocities are uncorrelated at a streamwise separation of half the length-scale L [14].

Table 5.3: Non-conforming embedded refinement in polar part of the unstructured grid of the pipe.

$Y^+ = u_\tau . y / v$	Rad. direction	Circumf. direction
$0 < Y^+ \leq 30$	4 cells	256 cells
$30 < Y^+ \leq 100$	4 cells	192 cells
$100 < Y^+ \leq 360$	8 cells	128 cells

An unstructured grid consisting of 740,000 cells is used to avoid having too many gridpoints in the core region of the pipe and in order to properly resolve the near-wall region. A polar grid is used for the first three layers with non-conforming embedded refinement as shown in Table **(5.3)**. Then the polar grid is made to match an octahedral bloc for the core region of the pipe (see Fig. **5.8**). The $2D$ grid is then extruded in the streamwise direction by 192 nodes. The first grid point near the pipe wall at which the axial velocity is computed is located at $Y^+ = 1.3$. 2 grid points are placed within the viscous sublayer, the depth of which approximately equals 5 wall units. A non-uniform grid is employed in the normal-to-the-wall direction within the circular part. This is done in order to locate more gridpoints in the near-wall region that is characterized by steep gradients and small energy-containing eddies. The Reynolds number of the simulation based on the pipe diameter D and on the centerline velocity u_c equals approximately 50,000 (based on bulk velocity u_b and shear stress velocity u_τ, it is 42,000 and 2,200 respectively).

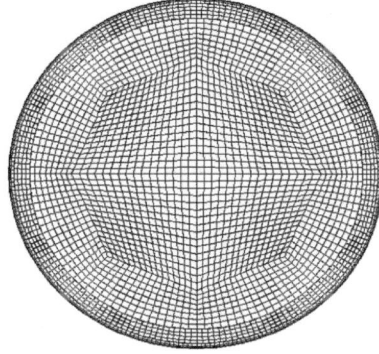

Figure 5.8: Unstructured grid used for LES

Periodic boundary conditions for velocity components are applied in the homogeneous streamwise axial direction. At the pipe wall, no-slip boundary conditions are imposed for all velocity components whereas Neumann boundary conditions are used for the pressure.

The filtered spatial and temporal evolution of an incompressible Newtonian fluid flow can be determined from the following equations:

$$\frac{\partial \bar{u}_i}{\partial x_i} = 0, \tag{5.2}$$

$$\frac{\partial \bar{u}_i}{\partial t} + \frac{\partial \bar{u}_i \bar{u}_i}{\partial x_j} = -\frac{\partial \bar{p}}{\partial x_i} + \frac{1}{Re} \frac{\partial^2 \bar{u}_i}{\partial x_j \partial x_j} + \frac{\partial \tau_{ij}}{\partial x_j}, \tag{5.3}$$

where,

$$\tau_{ij} = \bar{u}_i \bar{u}_j - \overline{u_i u_j}. \tag{5.4}$$

Here τ_{ij} is the subgrid scale (SGS) stress tensor. It is modeled using the algebraic eddy-viscosity model proposed by Smagorinsky [15]:

$$\tau_{ij} - \frac{1}{3}\delta_{ij}\,\tau_{kk} = -2\nu_{SGS}\bar{S}_{ij}, \tag{5.5}$$

With ν_{SGS} is the subgrid scale viscosity

$$\nu_{SGS} = (C_s \Delta)^2 |\bar{S}|. \tag{5.6}$$

Here C_s is a constant, and $|\bar{S}| = \left|2\bar{S}_{ij}\bar{S}_{ij}\right|^{1/2}$, where $\bar{S}_{ij} = \frac{1}{2}(\partial_j \bar{u}_i + \partial_i \bar{u}_j)$ is the resolved rate-of-strain tensor. The length scale Δ is taken equal to $2h$, where h is the grid spacing. The value of the constant C_s is taken equal to 0.065.

To drive the flow, the mean pressure gradient $-\nabla P$ has a non zero component only in the streamwise direction in order to balance the net viscous friction at the pipe wall. This non-zero component equals 4 in non-dimensional form when scaled with D and u_τ.

The time step Δt in the numerical simulations equals $0.03t^*$. t^* is the integral time scale defined as the ratio of pipe diameter D and velocity at the center of the pipe u_c. This time step is dictated by the numerical stability issues and the dispersed phase simulation. For numerical reasons linked to the solving of the system of stochastic differential equation (SDE) described in Chapter 4, the size of the time step has to be an order of magnitude smaller than the relaxation time of the smallest particles tracked for reasons discussed in Chapter 4. The LES computations are initiated from a randomly generated field with mean velocity and turbulent kinetic energy profiles fitted to analytical formulae. The time advancement was carried out until $t = 900t^*$ to achieve a flow field independent of the initial conditions. At $t = 900t^*$, the total shear stress profile versus the pipe radius showed an almost linear distribution, indicating that the computations had reached a nearly statistically steady state. From $t = 900t^*$, the calculations were continued until $t = 1100t^*$. In this interval, the final statistical data have been accumulated by spatial averaging in the streamwise and circumferential directions.

It is postulated that in practical applications of LES numerical dissipation will always be a significant part of the overall dissipation and it must be accounted for in any assessment of the quality of LES results. However, this assessment is not straightforward as it is in the case for RANS validation. This is mainly because of the well known difficulty of discriminating between the modeling errors and the numerical discretization errors that are both functions of the grid resolution [16].

Speziale [17] pointed that a reliable LES is the one that becomes a DNS when the grid resolutions is as small as the Kolmogorov scales. Consequently, one cannot seek a grid independent LES, as we usually do for RANS [18]. This is because a grid independent LES is essentially DNS, and the philosophy of LES that is based on grid dependency, loses its meaning. Celik *et al.* [19] developed a method to assess the quality of LES results. It consists of estimating an index of quality which is a measure of the percentage of the resolved turbulent kinetic energy. They stated that if more than 75% of the kinetic energy is captured (less than 25% of the turbulent kinetic energy is carried by the discarded SGS scales), then the LES is considered adequate. For the present LES, the ratio of the SGS kinetic energy to the total is given by Fig. (5.9) as function of the pipe radius. The SGS kinetic energy is estimated using Equations (4.5) and (4.6) of Chapter 4. As expected, more energy is filtered out near the wall, in particular in the buffer layer ($\approx 12\%$), compared to the core region of the pipe ($\approx 8\%$). These values might be underestimated because of some artificial dissipation that is introduced by the numerics and act along with the SGS dissipation.

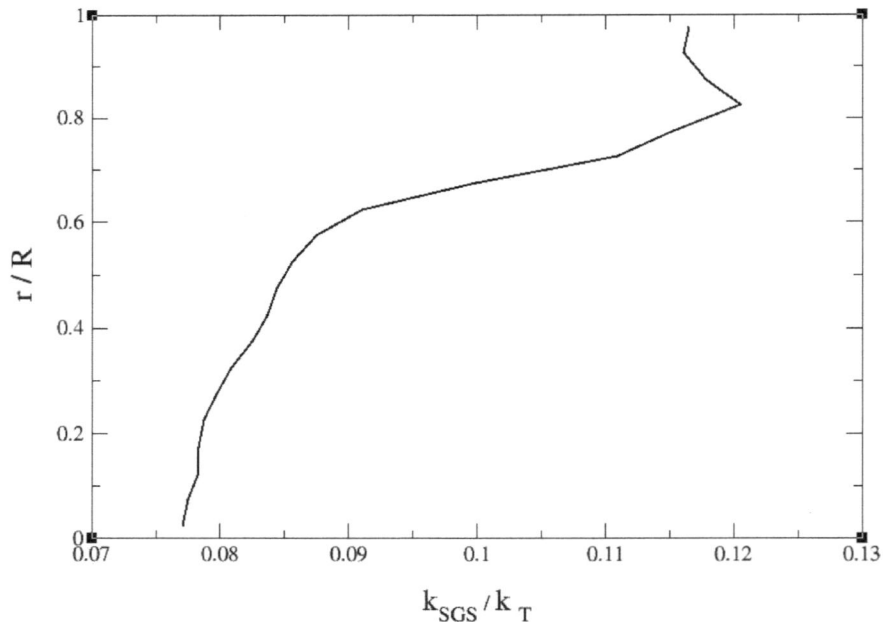

Figure 5.9: Ratio of the SGS kinetic energy k_{SGS} to the total turbulent kinetic energy k_T against the non-dimensionalied pipe radius (r/R). The total turbulent kinetic energy is the sum of the SGS kinetic energy and the LES-resolved kinetic energy $k_T = k_{SGS} + k$

Some works investigating numerical dissipation in LES have shown that the numerical errors may introduce a viscosity that is comparable to the SGS viscosity [19] when some upwinding is introduced. However, only a full central difference scheme is used in the current simulations.

LES calculations provide the large-scale velocity field and allow, under certain assumptions, the evaluation of the subfilter or SGS kinetic energy and its dissipation rate. These statistics are necessary for the construction of the velocity field seen by inertial particles, as it is explained in Chapter 4. It is well known that physical phenomena such as mixing, combustion and particle dispersion are strongly dependent on the intensity of turbulent fluctuations and the convection by these fluctuations. Hence in most of applications of LES, the prediction of turbulence statistics is at least as important as prediction of mean flow quantities.

Figs. **(5.10)** to **(5.12)** show mean axial velocity and rms turbulent fluctuations profiles (nondimensionalized by the friction velocity u_τ). Present LES results are found to compare well with mean flow properties and turbulence intensities of Laufer's experiment [20] and the LES of Uijttewaal *et al.* [21] (using the same number of cells and the dynamic procedure for the SGS model). These experimental and numerical results are all conducted at the same Reynolds number. For comparison we have also plotted the universal curves defined by $< w^+ >= Y^+$ for $Y^+ < 5$, and $< w^+ >= 2.5.\ln(Y^+) + 5.5$ for $Y^+ > 30$. Also comparison is made to more recent experimental works conducted by Den Toonder [22] ($Re_\tau = 1380$) and Zagarota [23]($Re_\tau = 2000$). Also analytical formulas for mean and turbulence statistics derived by Laurence [24] are presented. These formulas are derived from DNS and LES databases for turbulent channel flows. Channel flow statistics are expected to be very similar to the statistics of pipe flow in particular in the log-law region with some differences toward the center. RANS predictions are also presented. They are obtained using the second-moment model (RST) to close the time-averaged Navier-Stokes equations. As far as the mean axial velocity is concerned, the numerical results compared very well with Laufer's experimental observations. Uijttewaal's LES predictions are the closest to the experimental results in particular in the region $60 < Y^+ < 400$. This is due to the use of the dynamic procedure of the SGS model. In the near-wall region, surprisingly the present LES is performing better than Uijttewaal's LES though a conventional Smagorinsky model is used for the modeling of the SGS effects. This can be attributed to the resolution enhancement brought about by the embedding of many non-conforming layers as it is shown in Table **(5.3)**.

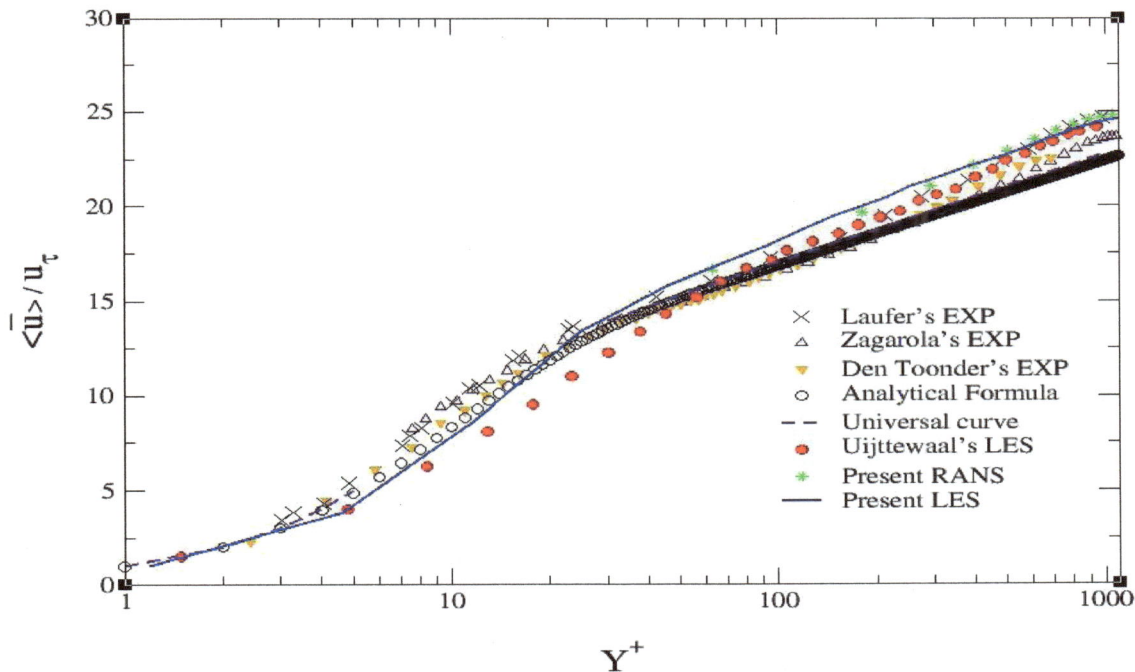

Figure 5.10: Mean axial velocity profile $< \overline{u} >$ non-dimensionalised with friction velocity u_τ versus radial distance in wall units Y^+.

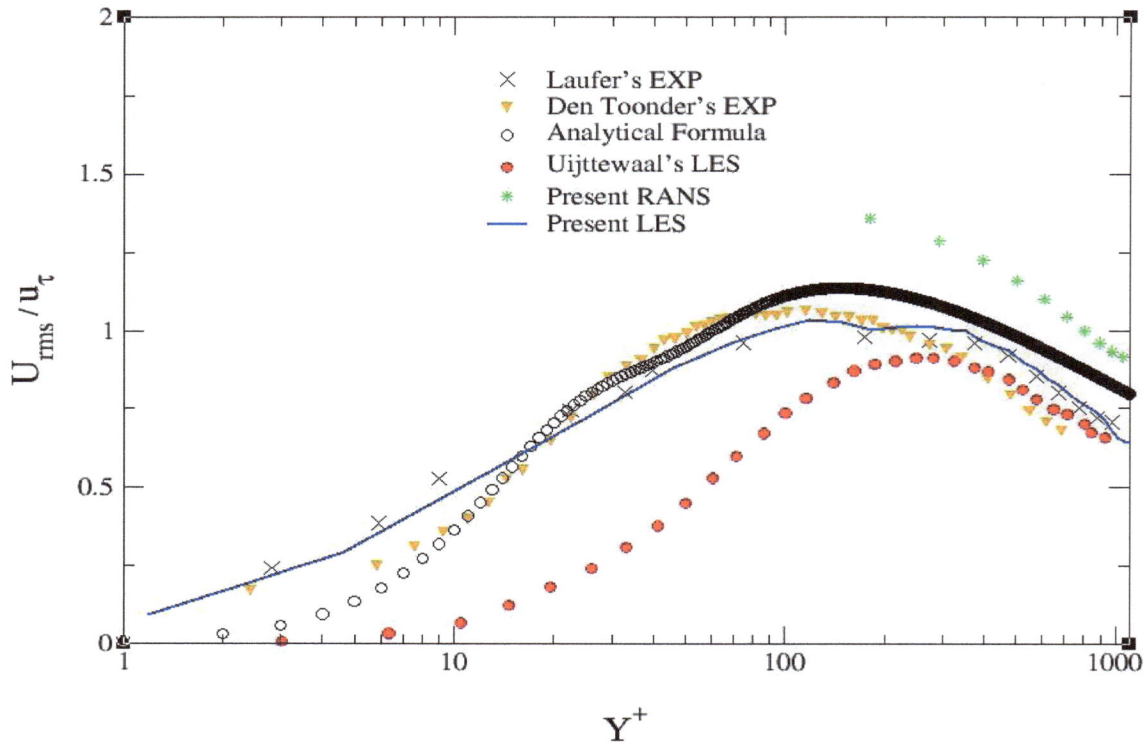

Figure 5.11: rms of the radial turbulent velocity fluctuations U_{rms} non-dimensionalised with friction velocity u_τ versus radial distance in wall units Y^+.

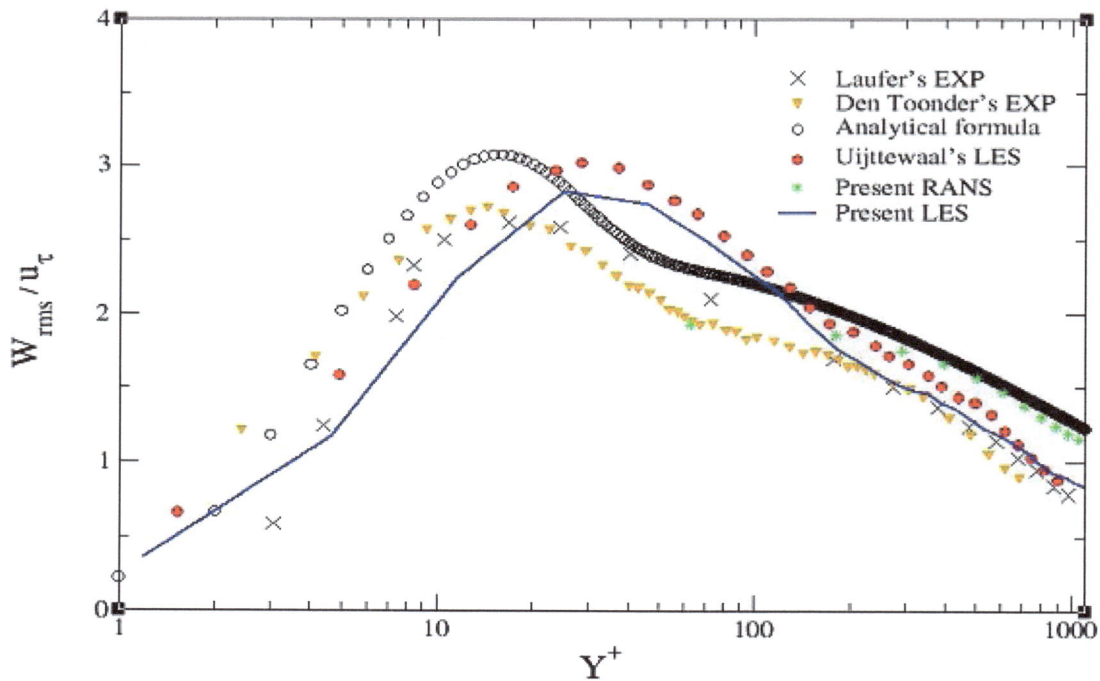

Figure 5.12: ms of the streamwise turbulent velocity fluctuations W_{rms} non-dimensionalised with friction velocity u_τ versus radial distance in wall units Y^+.

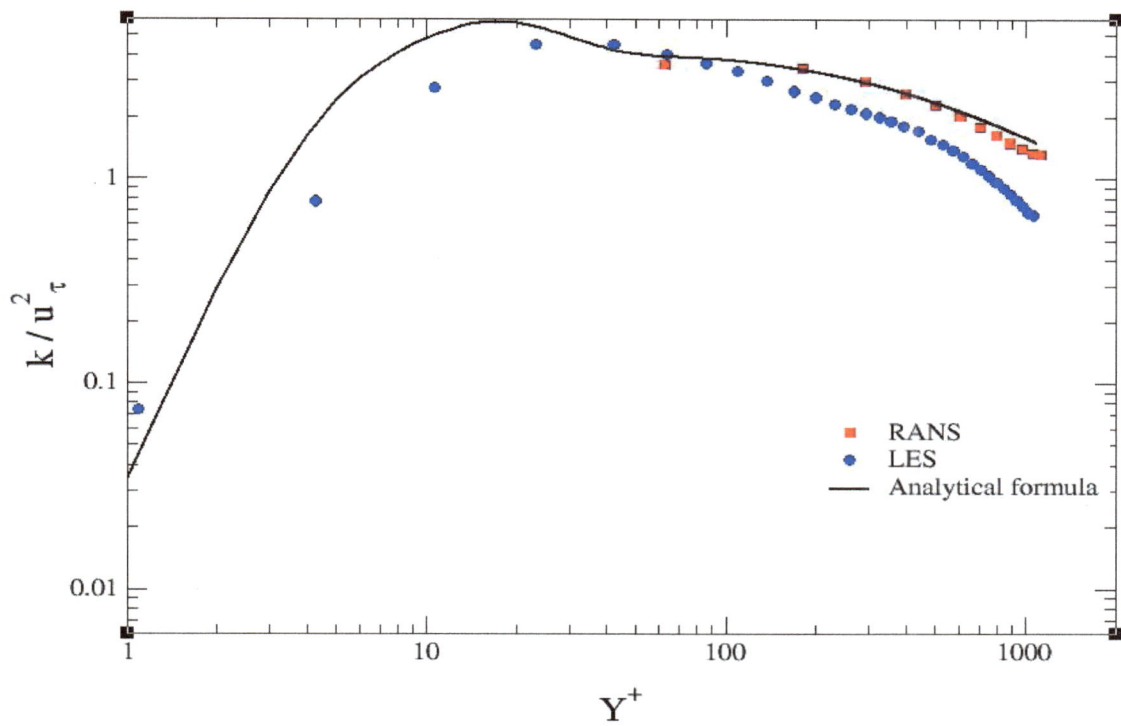

Figure 5.13: Turbulent kinetic energy k non-dimensionalised with friction velocity u_τ^2 versus radial distance in wall units Y^+.

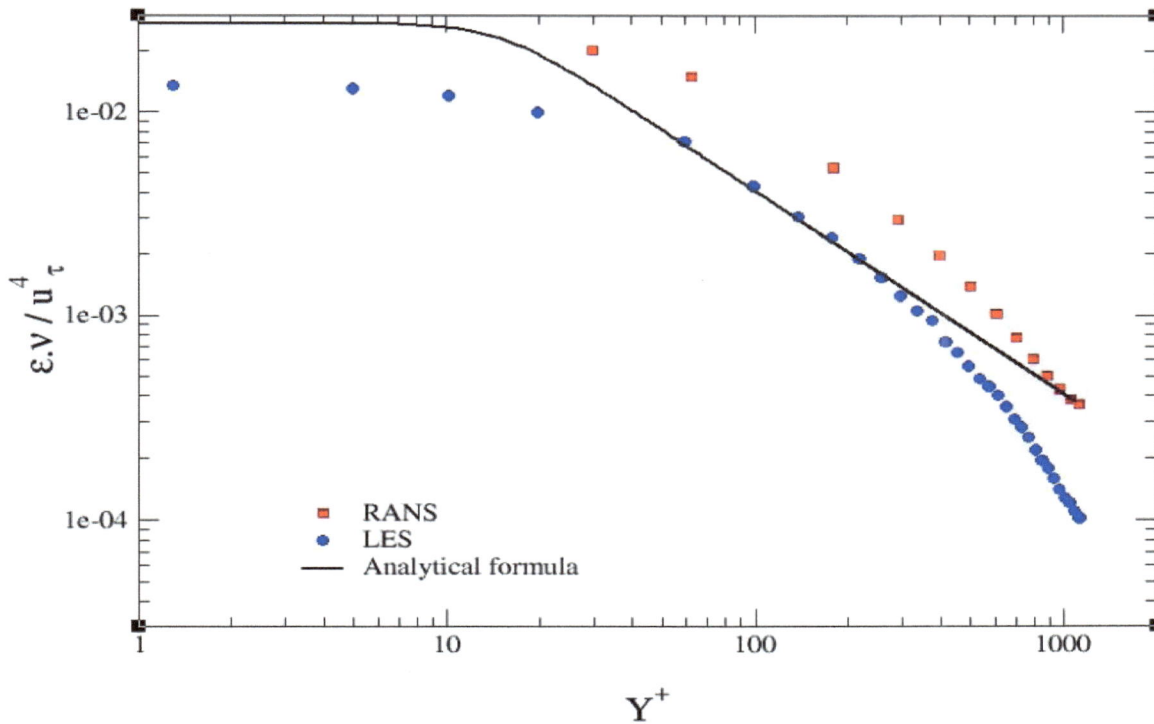

Figure 5.14: Dissipation rate of the turbulent kinetic ε non-dimensionalised with friction velocity u_τ and kinematic viscosity ν versus radial distance in wall units Y^+.

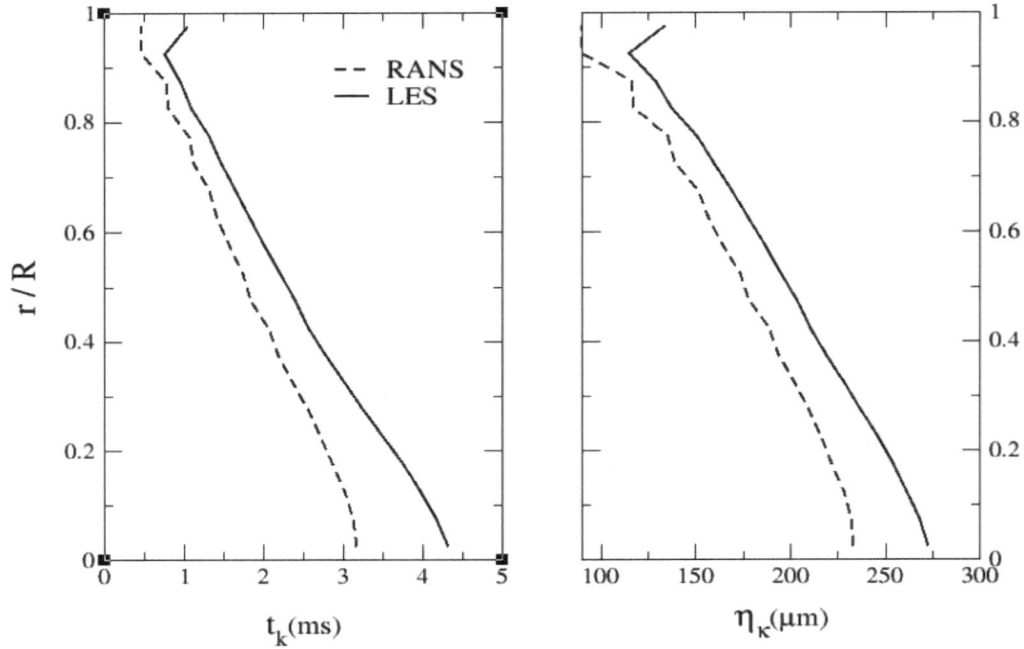

Figure 5.15: RANS and LES predictions of Kolmogorov time scale t_k and length scale η_k against the non_dimensionalised pipe radius (r/R) computed using Eqn. (5.7)

For numerical predictions of rms turbulent fluctuations, the superiority of LES results over RANS predictions are clear for both streamwise and radial rms fluctuations. Moreover, results of the present LES match the experimental results more closely than Uijttewaal's LES predictions. In particular near the wall and for the radial rms turbulent fluctuations that are responsible of particle radial dispersion.

In Figs. **(5.13)** and **(5.14)**, the turbulent kinetic energy (TKE) k and its rate of dissipation ε, as they are predicted by RANS, are compared to those obtained by LES. They are nondimensionalized by u_τ^2 and u_τ^4/ν respectively. The numerical results are also compared to the analytical formulae for the TKE and its rate of dissipation for channel flow [24]. Turbulent kinetic energy computed by RANS is higher up to 60% than the resolved turbulent kinetic energy provided by LES. This is true almost every where in the computation domain except in the near-wall region where the resolved TKE has a higher peak compared to the RANS computations. Same trend is noticed for the dissipation rate of TKE.

Accuracy in the prediction of both turbulent kinetic energy and its dissipation rate is important for the dispersed phase simulation. Indeed, they are used to compute some important physical quantities such as Kolmogorov time and length scales t_k and η_k as well as the Lagrangian and Eulerian time scales of the turbulent flow T_L and T_E. These physical quantities are important to predict the influence of the dispersive character of turbulence on solid particles of different response times τ_p. In fact, turbulent fluctuations present in any turbulent flow have time scales that range from the Kolmogorov time scale to the integral time scale. Particles travelling in this flow will only sense turbulent fluctuations whose time scales are larger than their response times. Also, the particle equation of motion is only valid for tracking particles with diameters that are an order of magnitude smaller than the Kolmogorov length scale of the turbulent flow. Consequently, predictions of such parameters are crucial for the simulation of dispersed flows.

Fig. **(5.15)** shows RANS and LES predictions of the Kolmogorov time and length scales. In Fig. **(5.16)**, the Lagrangian time scale is presented. The Kolmogorov time and length scales are computed as the following:

$$\eta_k = \left(\frac{\nu^3}{\varepsilon}\right)^{\frac{1}{4}}, t_k = \left(\frac{\nu}{\varepsilon}\right)^{1/2}, \tag{5.7}$$

where ν is the kinematic viscosity.

The Lagrangian time scale is computed according to the following formula [25]

$$T_L = C_T \cdot \frac{k}{\varepsilon}. \qquad (5.8)$$

The coefficient C_T is set to 0.3, which was obtained by Milojevic *et al.* [26]. They carried out a computer optimization of the experimental results of Snyder and Lumley [27]. The Eulerian time scale T_E is then deduced from the Lagrangian time scale according to the formula $T_E = T_L/\beta$ with $\beta = 0.356$ [9].

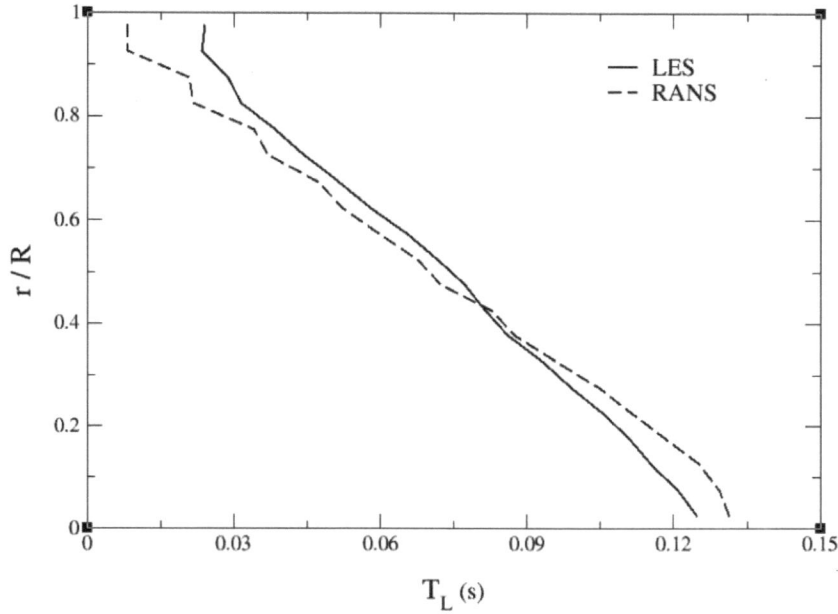

Figure 5.16: RANS and LES predictions of Lagrangian integral time scale of turbulence T_L against the adimensionalized pipe radius (r/R) computed using Eqn. (5.8)

For the prediction of these parameters, the discrepancies noticed between RANS and LES predictions are due mainly to the discrepancy in the prediction of the dissipation rate of TKE.

The discretization in the solver used to carry on simulations for applications studied in this book is based on the collocated finite-volume approach. It allows solving Navier–Stokes and scalar equations on hybrid and non-conform unstructured grids. Velocity and pressure coupling is ensured by the prediction/correction method with a SIMPLEC algorithm. The collocated discretization requires the Rhie and Chow interpolation in the correction step to avoid oscillatory solutions. A second order centred scheme (in space and time) is used. The flow solver has been extensively tested for LES of single-phase flows [28].

5.2.2 Particle Phase

From a point source located at the center of the pipe, solid particles are released and tracked into the turbulent flow described in the previous section. The physical properties of these solid particles are summarized in Table (5.2). The simplest way to characterize the dynamics of particle motion is by means of its Stokes relaxation or response time and the corresponding Stokes number given by:

$$St = \frac{\tau_p}{T_E}, \text{ where } \tau_p = \frac{\rho_p d_p^2}{18\mu}. \qquad (5.9)$$

Here ρ_p is the mass density of the particles, d_p is the diameter of the particles, μ is the dynamic viscosity of the fluid, and T_E is the Eulerian time scale of the fluid phase.

Since large particles are also considered in this study, the particle Reynolds number is expected to often exceed unity as it is shown on Fig. **(5.17)**. A non-linear drag coefficient described by Equation (5.11), taking into account the high particle Reynolds number, is therefore more appropriate. As a consequence the actual particle response time will, at high particle Reynolds numbers, be smaller than the one defined by Equation (5.9). This fact is depicted in Fig. **(5.18)**. Numerical predictions of the particle Reynolds number in Fig. **(5.17)** show that the Reynolds number of $37\mu m$ and $57\mu m$ particles is well beyond unity irrespective of the particle position in the flow. It increases as the particles move towards the wall. The same tendency is also noticed for the $5\mu m$ particle though the flow around them is in the Stokesian regime characterised by a particle Reynolds number less than unity.

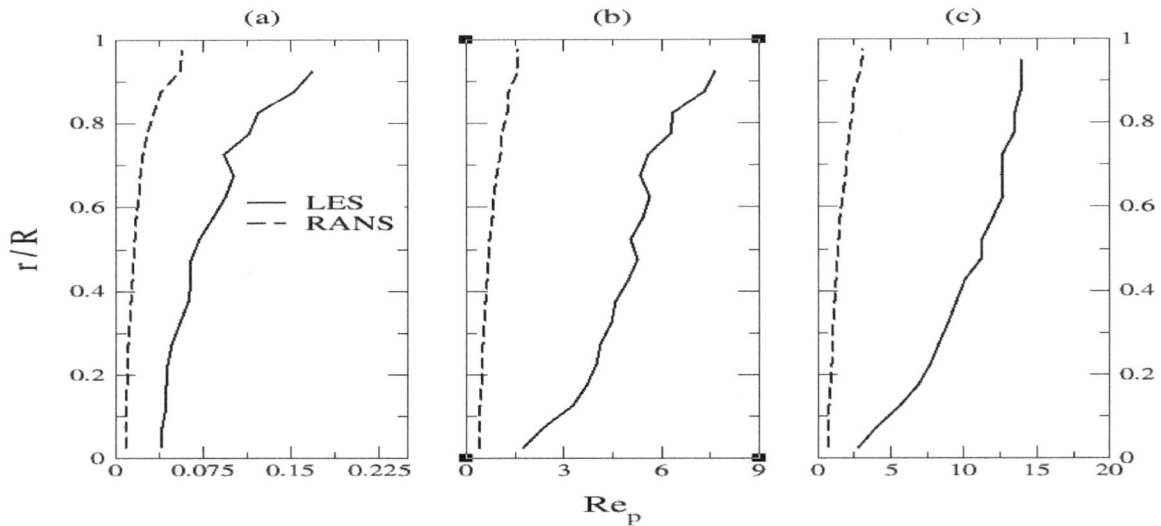

Figure 5.17: RANS and LES predictions of particle Reynolds number $\mathbf{Re_p}$ based on mean slip velocity against the non-dimensionalised pipe radius $(\mathbf{r/R})$. (a) $\mathbf{5\mu m}$ particles, (b) $\mathbf{37\mu m}$ particles, (c) $\mathbf{57\mu m}$ particles.

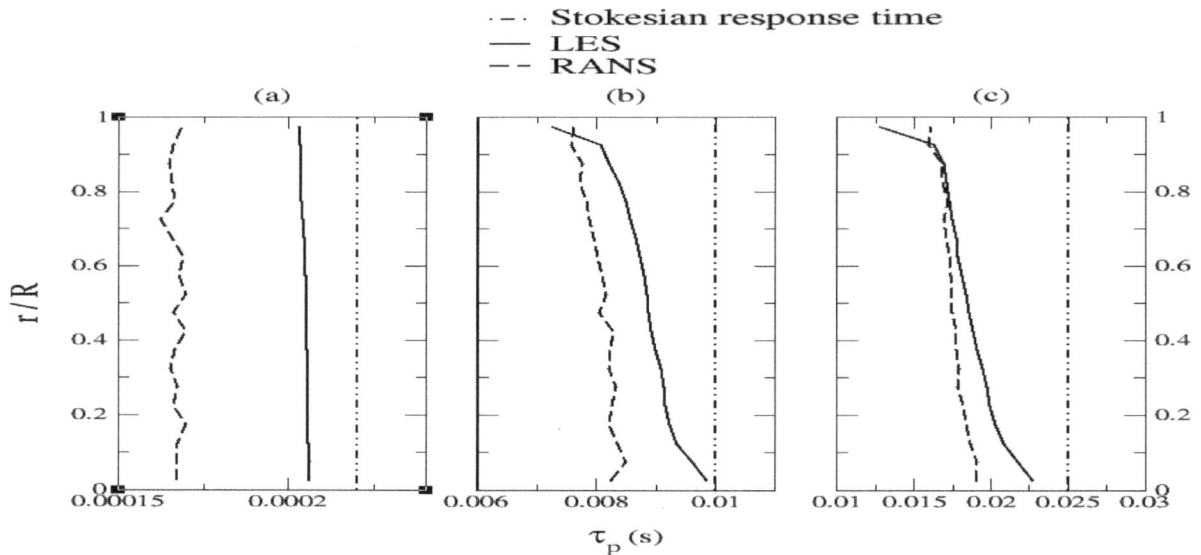

Figure 5.18: RANS and LES predictions of particle response time $\mathbf{\tau_p}$ (Eqn. (5.10)) compared to the Stokesian response time (Eqn. (6.9)). (a) $\mathbf{5\mu m}$ particles, (b) $\mathbf{37\mu m}$ particles, (c) $\mathbf{57\mu m}$ particles.

Injection conditions concerning the position and initial velocity of particles were not documented in the experimental work. For that reason, solid particle velocity is set equal to that of the filtered fluid velocity at the position of the particle. The same value also is given to the seen fluid velocity. The initial particle position is chosen randomly on a cross section of 3 mm in diameter that corresponds to the injection tube diameter. For RANS, since only mean velocity is available, the particle velocity at the injection is set equal to the mean velocity plus a random fluctuation based on the turbulent kinetic energy and a random number generated using a Gaussian distribution. As for LES, the seen fluid velocity is set equal to the particle velocity at the injection section.

The basic concept of Lagrangian particle tracking is to integrate the equations of motion for individual particles. The majority of the calculations found in the literature are based on the Maxey and Riley equation (MR) [29]. As a result of the high density ratio between particle and fluid densities, this equation becomes reasonably simple and only the drag and gravity forces will be retained since other forces are in this case negligible. As a prerequisite for the use of MR equation, the particle diameter has to be smaller than the Kolmogorov length scale. As it is shown in Fig. **(5.15)**, both RANS and LES results predicted a varying Kolmogorov length scale with the pipe radius. It is found around $250\mu m$ near the pipe center and decreases to reach approximately $100\mu m$ near the wall. The largest particle diameter considered in this work is $57\mu m$ which justifies the use of the MR equation.

Therefore the tracking of the solid particles within the turbulent flow obeys the stochastic differential systems described in Chapter 4. (system (4.3) for the standard formulation or system (4.15) if the complete formulation is used) with :

$$\tau_p = \frac{\rho_p}{\rho_f} \frac{4d_p}{3C_D|u_s - u_p|} \tag{5.10}$$

$$C_D = \begin{cases} \frac{24}{Re_p}\left(1 + 0.15Re_p^{0.687}\right) \ if \ Re_p < 1000 \\ 0.44 \ if \ Re_p > 1000. \end{cases} \tag{5.11}$$

Since it is extremely rare that solid particle position would coincide with grid points, a tri-linear interpolation scheme is used to obtain the velocities between the grid points.

The interaction of solid particles with the wall is considered perfectly elastic. According to the Sommerfeld criteria [25], the turbulence level and the size of inertial particles considered in Arnason's experiment and in the numerical simulations do not give rise to a wall-collision dominated flow. Indeed, in many practical situations, the confinement may considerably influence the particle motion and hence the velocity statistics as a result of particle-wall collisions. The near-wall region is important regarding particle deposition since turbulent fluctuations in radial direction are responsible for the radial particle transport [25]. The core region turbulence away from the wall, covering greater part of the cross-section, affects more the overall turbulent dispersion of particles. In this region gradients are small and the turbulence exhibits a more or less homogeneous and isotropic behavior [21]. Based on some physical and dimensional considerations, Sommerfeld [25] formulated an expression for the minimum particle diameter that can give rise to a wall-dominated flow for a given level of turbulence, it reads:

$$d_p \approx \sqrt{\frac{18\mu D}{0.7K\rho_p <u_p>}}. \tag{5.12}$$

Where $< u_p >$ is the mean axial particle velocity, K is the ratio between the transverse velocity fluctuation v_p' and $< u_p >$.

Sommerfeld [25] considered that a collision-dominated two-phase flow can take place when more than 30% of the particles collide with the wall before these respond to the fluid flow. It is obvious from Sommerfeld's formula that the minimum particle size for which the particle motion is strongly determined by wall collision decreases with increasing transverse velocity fluctuations and increases with the pipe diameter, since the particles have more time to respond to the fluid flow. For the turbulent case in hand, the minimum particle diameter for which the flow can be dominated by wall collision is computed using Equation (5.12). It turned out to be: $d_p = 108\mu m$. The largest diameter tracked in this test case is: $d_p = 57\mu m$. This fully justifies the use of a simple law of wall-collision such as the one describing an elastic rebounce. Dilute concentration of particles is considered for all the simulations (volume fraction $\alpha_p < 10^{-6}$).

Solid particles are injected at time $t = 1100t^*$ and calculations are advanced in time using the same time step $\Delta t = 0.03t^*$ used for the single-phase computations. After a time $t = 100t^*$, the number of particles in the flow domain becomes stationary and therefore collection of particle statistics starts. This lasts for another time equals to $t = 100t^*$. In this interval, the final statistical data have been accumulated by spatial averaging in circumferential direction for each pipe section considered. In fact, a time $t = 100t^*$ is enough to allow solid particles a few response times to adjust to the surrounding fluid, before their statistics are collected. For the heaviest particles considered in these simulations, $t = 100t^* = 40.\tau_{p,57}$.

5.3 RESULTS AND DISCUSSION

The numerical predictions of Arnason's experiment using RANS and LES were collected at three sections downstream the seeding point ($/D$ = 3.5, 5.6 and 7.5). They consist of particle concentration, streamwise and radial velocity profiles for two different particle diameters, namely $5\mu m$ and $57\mu m$ particles. For the $37\mu m$ particles, only concentration and radial velocity profiles at two sections (Z/D =3.5 and 5.6) are provided. Using these results, the dispersion coefficient ϵ_p for each particle size is evaluated and then compared to the experimental measurements. Averaging is made over the number of particles present in a cell at a given time and a given position and then in the circumferential directions.

Figs. (**5.19**) to (**5.21**) display numerical predictions of the $5\mu m$ particle statistics. They consist of concentration, streamwise and radial velocity profiles computed using RANS, LES with and without taking into account the subfilter effects on particles dispersion. As expected, a significant improvement in the results was made possible by introducing the effects of the subfilter scales on the motion of the $5\mu m$ particles. These particles have Stokes numbers less than one (evaluation of the Stokes numbers based on SGS time scale for the $5\mu m$ particles shows that they vary between 0.006 and 0.025) as it is shown in Fig. (**5.28**). This class of particles senses the turbulent fluctuations associated with SGS scales irrespective of their positions in the flow domain; therefore inclusion of such SGS fluctuations is crucial. It is clear that LES predictions closely match the experimental results, though slightly underestimating the concentration profiles near the wall. RANS predictions increasingly overestimate the particle concentration spread as we move downstream. The LES without subfilter model shows an initial underestimation of the spread, and this lag remains constant downstream. With the subfilter model incorporated in the LES, this initial defect is now well corrected.

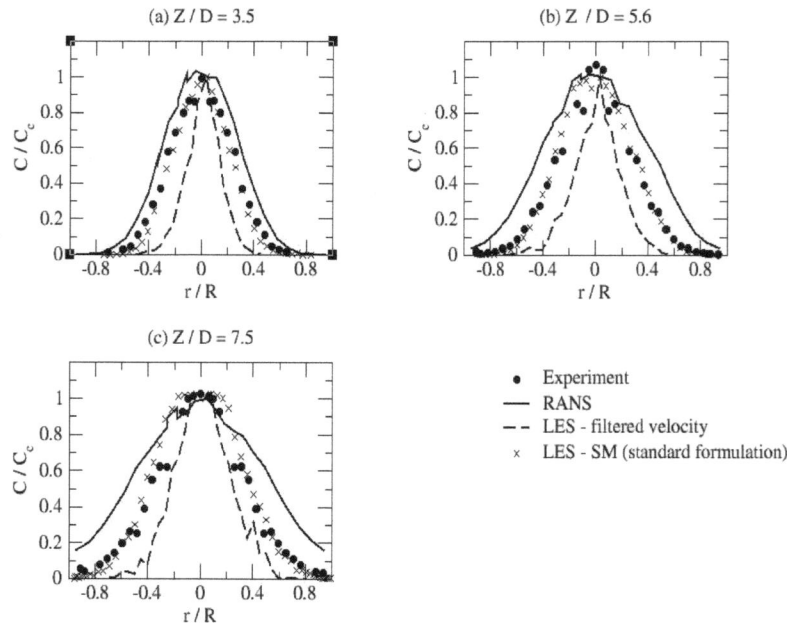

Figure 5.19: Concentration profiles C non-dimensionalized by the concentration at the pipe center C_c of **5μm** particles versu the non-dimensionalised pipe radius (**r/R**) at different locations (**Z/D**) downstream the injection point. (a) **Z/D** = 3.5, (d) **Z/D** = 5.6, (c) **Z/D** = 7.5. Predictions of RANS, LES, using only the filtered velocity and LES using the standard formulation of the stochastic model are compared to the experimental observations.

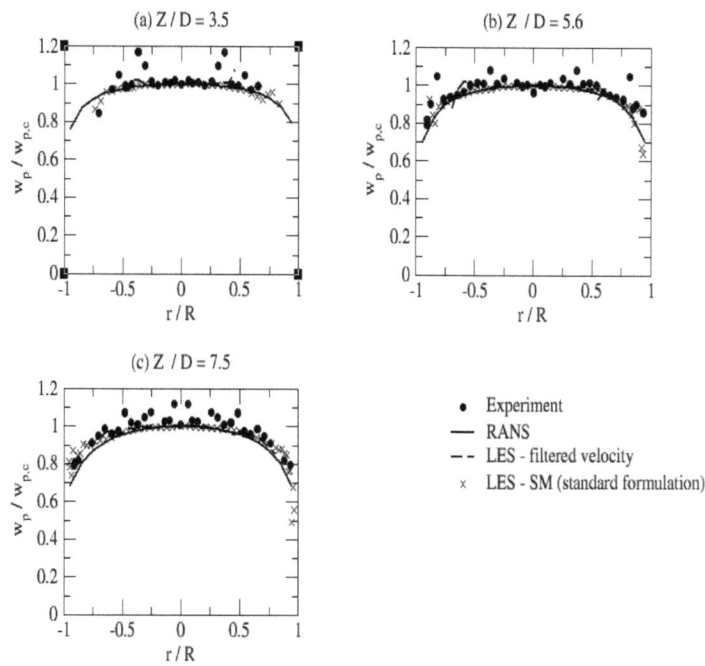

Figure 5.20: Streamwise velocity profiles w_p non-dimensionalised by the streamwise velocity at the pipe center $w_{p,c}$ of **5μm** particles versus the non-dimensionalised pipe radius at different locations (Z/D) downstream the injection point. (a) Z/D = 3.5, (d) Z/D = 5.6, (c) Z/D = 7.5 predictions of RANS, LES using only the filtered velocity and LES using the standard formulation of the stochastic model are compared to the experimental observations.

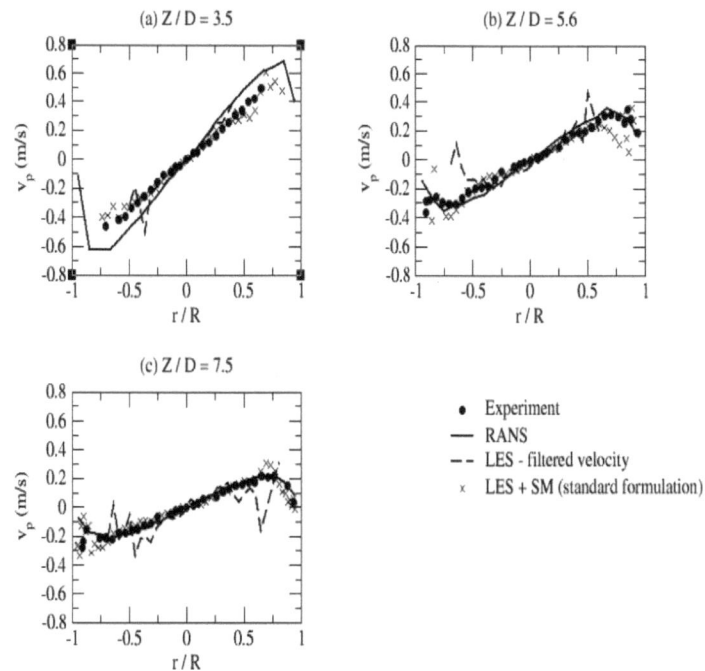

Figure 5.21: Radial velocity profiles v_p of **5μm** particles versus the non-dimensionalised pipe radius at different locations (Z/D) downstream the injection point. (a) Z/D = 3.5, (d) Z/D = 5.6, (c) Z/D = 7.5. Predictions of RANS, LES using only the filtered velocity and LES using the standard formulation of the stochastic model are compared to the experimental observations.

Figs. **(5.22)** and **(5.23)** display numerical predictions of the $37\mu m$ particle statistics. They consist of concentration and radial velocity profiles computed using RANS, LES with and without taking into account the subfilter effects on particles dispersion. It is clear from Fig. **(5.28)** that the influence of the SGS fluctuations is dependent on the $37\mu m$ particle position in the flow domain (evaluation of the Stokes numbers based on SGS time scale for the $37\mu m$ particles shows that they vary between 0.5 in the core region and 2.5 near the wall). The $37\mu m$ particles sense less and less SGS fluctuations as they move toward the wall. Also they do sense only a portion of the SGS fluctuations which is witnessed by their relatively high Stokes numbers compared to the ones of the $5\mu m$ particles. This has resulted in a relatively small difference on the particle statistics between LES without and with subfilter fluctuations modeling. Results of LES with the stochastic modeling compare very well with the experimental observations. RANS predictions are also in good agreement with the experiment especially for the concentration profiles but they do overestimate the radial velocity.

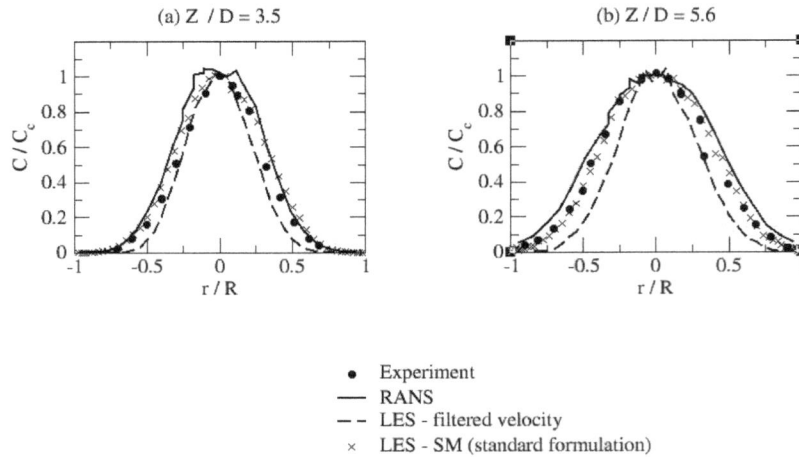

Figure 5.22: Concentration profiles C non-dimensionalized by the concentration at the pipe center C_c of $37\mu m$ particles versus the non-dimensionalised pipe radius at different locations (Z/D) downstream the injection point. (a) Z/D = 3.5, (d) Z/D = 5.6, (c) Z/D = 7.5. Predictions of RANS, LES using only the filtered velocity and LES using the standard formulation of the stochastic model are compared to the experimental observations.

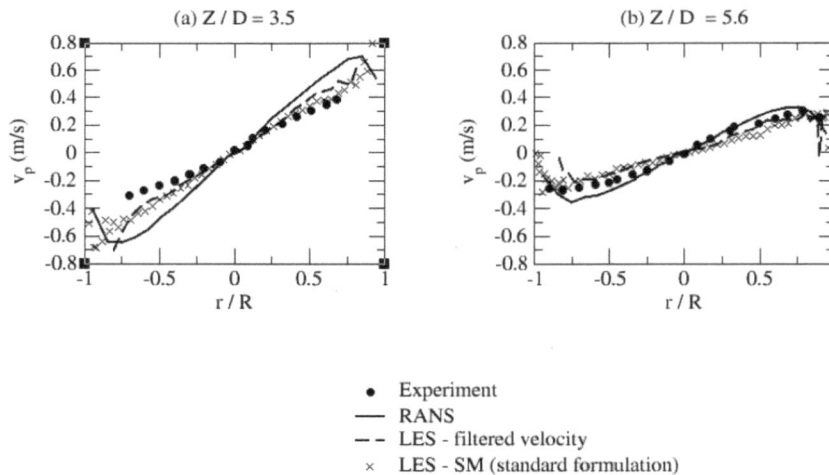

Figure 5.23: Radial velocity profiles v_p of $37\mu m$ particles versus the non-dimensionalised pipe radius at different locations (Z/D) downstream the injection point. (a) Z/D = 3.5, (d) Z/D = 5.6, (c) Z/D = 7.5. Predictions of RANS, LES using only the filtered velocity and LES using the standard formulation of the stochastic model are compared to the experimental observations.

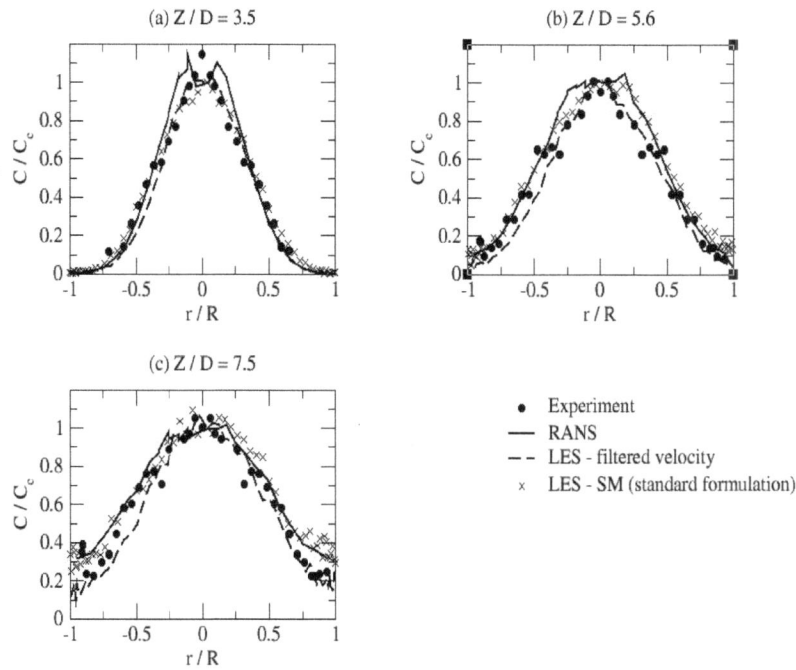

Figure 5.24: Concentration profiles C non-dimensionalized by the concentration at the pipe center C_c of **57μm** particles versu the non-dimensionalised pipe radius (r/R) at different locations (Z/D) downstream the injection point. (a) Z/D = 3.5, (d) Z/D = 5.6, (c) Z/D = 7.5. Predictions of RANS, LES, using only the filtered velocity and LES using the standard formulation of the stochastic model are compared to the experimental observations.

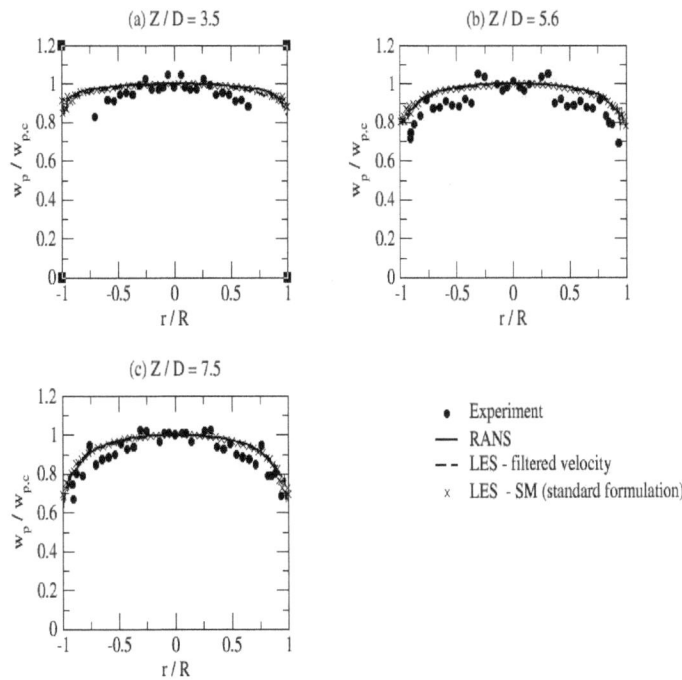

Figure 5.25: Streamwise velocity profiles w_p non-dimensionalised by the streamwise velocity at the pipe center $w_{p,c}$ of **57μm** particles versus the non-dimensionalised pipe radius at different locations (Z/D) downstream the injection point. (a) Z/D = 3.5, (d) Z/D = 5.6, (c) Z/D = 7.5 predictions of RANS, LES using only the filtered velocity and LES using the standard formulation of the stochastic model are compared to the experimental observations.

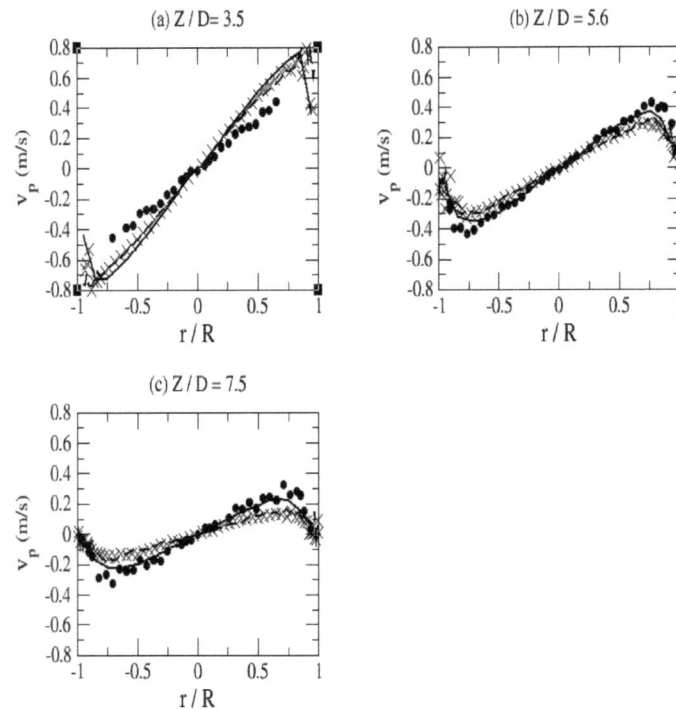

Figure 5.26: Radial velocity profiles v_p of **57μm** particles versus the non-dimensionalized pipe radius at different locations (Z/D) downstream the injection point. (a) Z/D = 3.5, (d) Z/D = 5.6, (c) Z/D = 7.5. Predictions of RANS, LES using only the filtered velocity and LES using the standard formulation of the stochastic model are compared to the experimental observations.

For the 57μm particles, Figs. **(5.24)** to **(5.26)** show that the effect of the subfilter scales on their motion is clearly very small. Their Stokes numbers based on the subfilter scale time scale exceed one (they are between 1 and 6 throughout the computational domain) as it is shown in Fig. **(5.28)**. Therefore they do not respond to the higher frequency subfilter scale motions. It was expected that there would be little effect of the subfilter scales on 57μm particles, which is verified on particle radial velocity profiles. However, a very small difference has been noticed on the particle concentration profiles, which is probably due to an excessive filtering out of the kinetic energy that is not well captured by the model. Both RANS and LES predict reasonably well the experimental results, with the LES results being marginally closer to the experiment for the concentration profiles.

It is worth mentioning that these numerical results concerning particle concentration, streamwise and radial velocity profiles are obtained using the *standard formulation* of the Langevin equation as it is described in Chapter 4.

In Fig. **(5.27)**, the influence of inertia and cross trajectory effects on the SGS time scale seen by solid particles is shown. As expected, Inertia and cross trajectory effects play an important role in defining the time scale with which a solid particle would see the fluctuations of the turbulent flow. Inertia results in an increase in the seen time scale whereas cross trajectory effects due to the presence of gravity causes a decrease in time scales seen by solid particles. This is mainly due to a loss of memory caused by a constant drift of particles through eddies of different sizes and turbulence levels. As explained in the introduction of this chapter, an increase in inertia results in an increase in particle time scale and a decrease in particle fluctuating velocity through the CT effects. The effects of these two competing parameters increase with the particle diameter. It is clear from both Figs. **(5.27)** and **(5.28)** that failing in accounting for inertia, results in the solid particle responding to less turbulence than it should. Neglecting the cross trajectory effects causes the solid particle to sense more turbulence than it should. It has to be emphasized that we have explained the effects of inertia and cross trajectory on the solid particle behavior towards the subgrid scale turbulence. The effect of these parameters on solid particle dispersion due to the whole turbulence fluctuations is investigated later in this chapter.

—— Lagrangian SGS time scale $T_{L,SGS}$

–– Lagrangian SGS time scale seen (Inertia effect) T^{*}_{SGS}

·–· SGS time scale seen (Inertia + CT effects) T^{*}_{SGS}

···· Kolmogorov time scale t_{k}

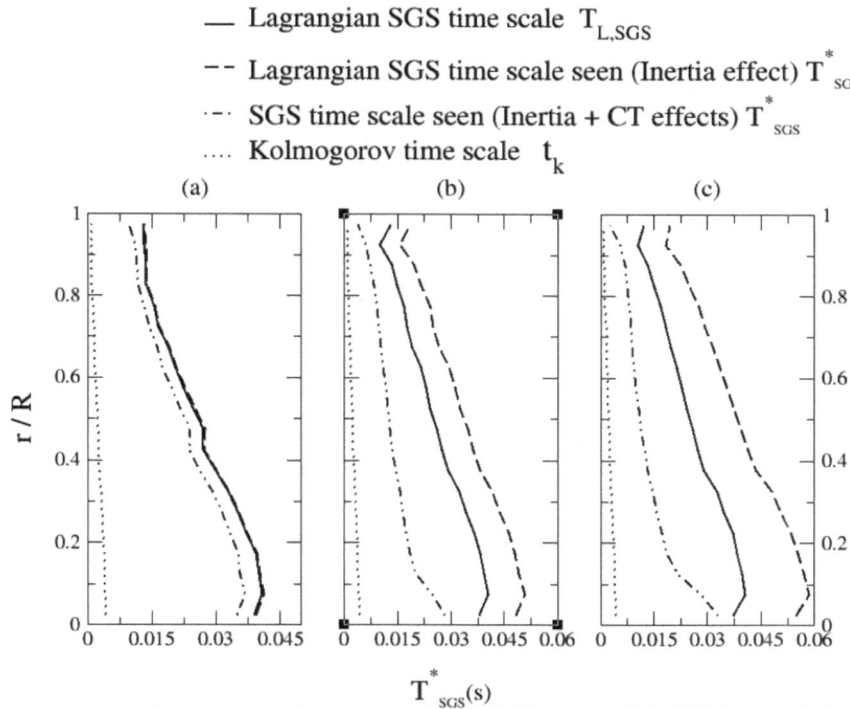

Figure 5.27: Lagrangian SGS time scale T^{*}_{SGS} with which particles see the SGS turbulence versus the non-dimensionalised pipe radius. Inertia and cross trajectory effects (CT) are included using Eqns (4.7) and (4.9). Kolmogorov time scale t_{k} presented also for comparison. (a) **5μm** particles, (b) **37μm** particles, (c) **57μm** particles.

—— Stokes number based on Lagrangian SGS time scale

–– Stokes number based on Lagrangian SGS time scale seen (Inertia effect)

·–· Stokes number based on SGS time scale seen (Inertia + CT effects)

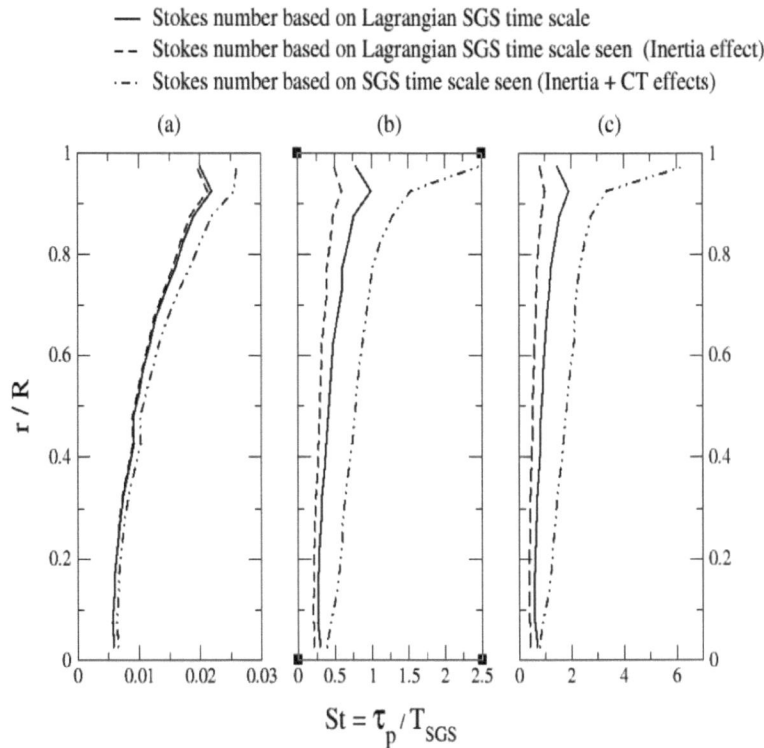

Figure 5.28: Particle Strokes number based on Lagrangian SGS time scale T_{SGS} versu the non-dimensionalised pipe radius. (a) **5μm** particles, (b) **37μm** particles, (c) **57μm** particles,.

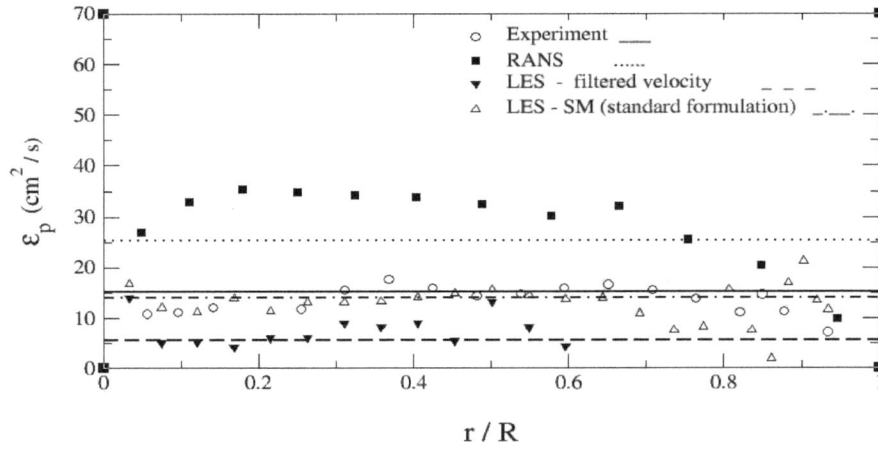

Figure 5.29: Long-time particle dispersion coefficient ϵ_p for **5μm** particles versus the non-dimensionalised pipe radius. Predictions of RANS, LES using only the filtered velocity and LES using the standard formulation of the stochastic model are compared to the experimental observations. Different lines are results computed with Eqn. (5.13). Different shapes of dots are results computed with Eqn. (5.14).

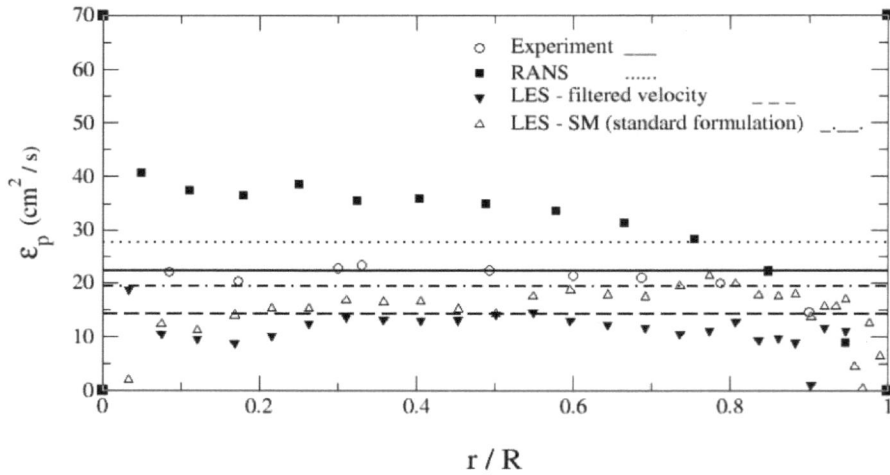

Figure 5.30: Long-time particle dispersion coefficient ϵ_p for **37μm** particles versus the non-dimensionalised pipe radius. Predictions of RANS, LES using only the filtered velocity and LES using the standard formulation of the stochastic model are compared to the experimental observations. Different lines are results computed with Eqn. (5.13). Different shapes of dots are results computed with Eqn. (5.14).

Figs. (5.29) to (5.31) show, for both numerical simulations, the prediction of the particle dispersion coefficient along with the experimental observations for the different particle sizes. Previous dispersion coefficient estimates were all based on observations of number density profiles of particles dispersing from a line or point source [2]. Then the dispersion coefficient or turbulent diffusivity of particles was related to the time rate of change of the mean square dispersion y^2 [10, 30]. It is given as:

$$\epsilon_p = \frac{1}{2}\frac{d}{dt}y^2. \tag{5.13}$$

This method is only correct if ϵ_p is constant throughout the flow field and the convective velocity is uniform. This is the case for homogeneous and isotropic turbulence. It is less meaningful for turbulent shear flows, however, which is the case of turbulent pipe flow.

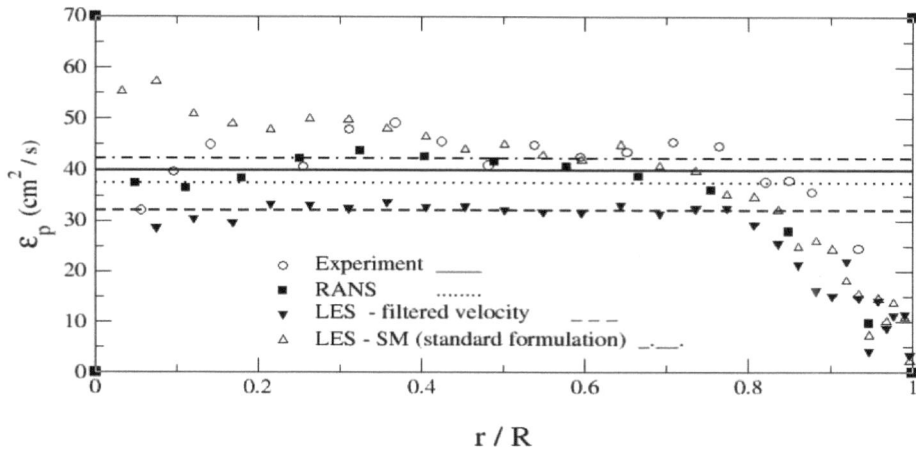

Figure 5.31: Long-time particle dispersion coefficient ϵ_p for **57μm** particles versus the non-dimensionalised pipe radius. Predictions of RANS, LES using only the filtered velocity and LES using the standard formulation of the stochastic model are compared to the experimental observations. Different lines are results computed with Eqn. (5.13). Different shapes of dots are results computed with Eqn. (5.14).

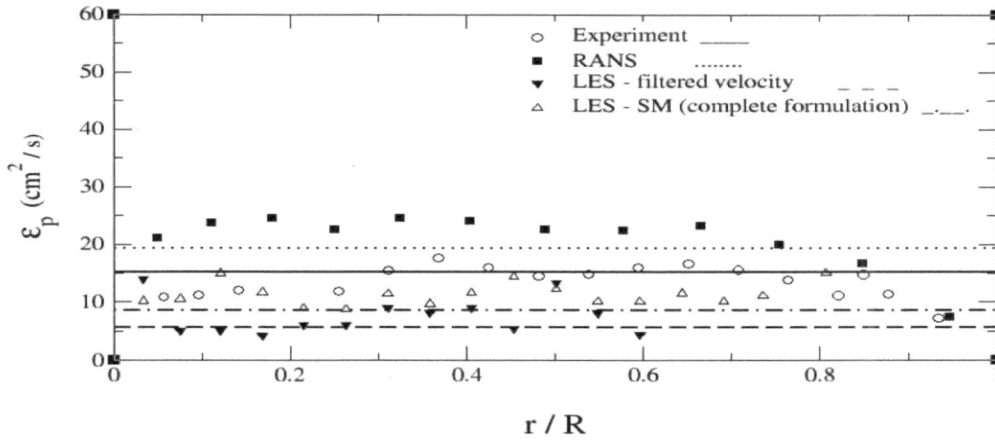

Figure 5.32: Long-time particle dispersion coefficient ϵ_p for **5μm** particles versus the non-dimensionalised pipe radius. Predictions of RANS, LES using only the filtered velocity and LES using the standard formulation of the stochastic model are compared to the experimental observations. Different lines are results computed with Eqn. (5.13). Different shapes of dots are results computed with Eqn (5.14).

A method is developed by Arnason and Stock [31] to allow local estimates of the diffusivity to be obtained when the turbulence is neither homogeneous nor isotropic. If the flux is considered to be caused only by gradient diffusion (Fick's law), the particle dispersion coefficient at every radial location can be computed according to the following formula, except in regions where $\partial C / \partial r$ is near 0:

$$\epsilon_p = -\bar{v}_p . \, C \left(\frac{\partial c}{\partial r} \right)^{-1}. \tag{5.14}$$

This allows the particle diffusivity to be computed locally at all measuring points. The required measurements are the average particle velocity $\bar{v}_p$ in the radial direction and the concentration profile C. Numerical predictions show that both RANS and LES predict reasonably well the particle dispersion coefficient of 57μm particles, whereas RANS predictions failed to match the experimental results of the 5μm and 37μm particles. LES predictions for the 5μm particles are excellent and highlight once again the importance of including the subfilter effects on small particles dispersion. These above-displayed numerical results are obtained using the standard formulation of the Langevin model described in Chapter (4).

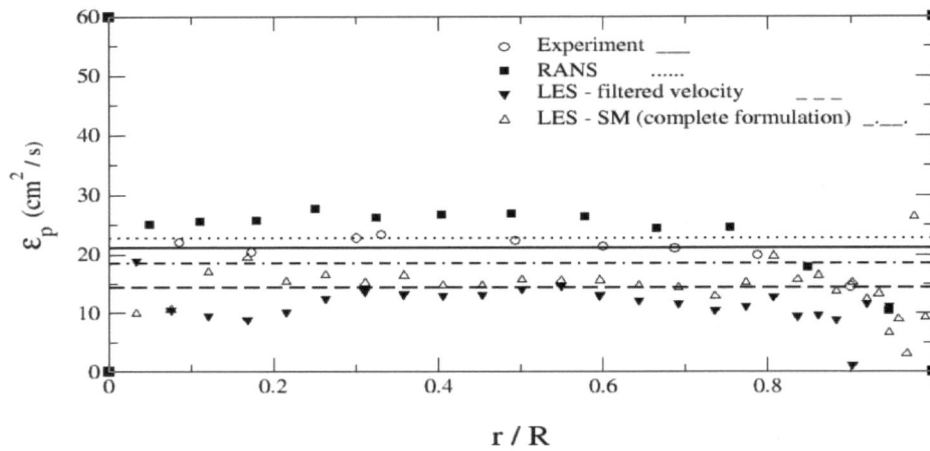

Figure 5.33: Long-time particle dispersion coefficient ϵ_p for **37μm** particles versus the non-dimensionalised pipe radius. Predictions of RANS, LES using only the filtered velocity and LES using the standard formulation of the stochastic model are compared to the experimental observations. Different lines are results computed with Eqn. (5.13). Different shapes of dots are results computed with Eqn. (5.14)

Like the experimental predictions, LES results show that for this case the particle dispersion coefficient increases with the particle diameter. RANS calculations predict almost the same dispersion coefficient for all the particle sizes considered. This is likely to be the result of an excessive turbulence fluctuations generated by the stochastic model for RANS. This stochastic model is based on the statistics of turbulence predicted by the RANS closure for the continuous phase and may fail if these statistics are not accurately predicted. Also, the Lagrangian stochastic model for RANS applications is designed with the aim to reproduce the whole spectrum of turbulence that an inertial particle should see. This turbulence is, in particular in the case of non-equilibrium flows, highly anisotropic and characterized by a wide range of length and time scales. Modeling accurately the intricate complexities of such turbulence is often problematic. This could explain the limitations faced by the Lagrangian stochastic modeling in the context of RANS. The picture becomes more optimistic when it comes to the use of the Lagrangian stochastic modeling to account for inertial particle transport by the subgrid motion for LES. Indeed the temporally and spatially narrow range of the small scales that are discarded by the filtering in LES are deemed to be quasi isotropic and homogeneous. Thus, the model is expected to reproduce them with a higher degree of accuracy.

The complete formulation of the Langevin equation is also tested and the results of the dispersion coefficients for the different particles are displayed in Figs. **(5.32)** to **(5.34)**. For RANS results concerning the 5μm and 37μm particles, the complete formulation of the Langevin equation has brought about some improvement while it resulted in an increased discrepancy concerning the dispersion of 57μm. As far as LES is concerned, the difference between the two formulations' results is only noticeable for the 57μm particles.

As it is discussed in Chapter 4, the parameter β that links the Eulerian time scale to the Lagrangian time scale is of great importance since it permits to use the results of the Eulerian field for the Lagrangian tracking of particles. Thus its influence on the performance of the model needs to be investigated. Three values of β are tested, namely 0.356 [9], 0.8 [32] and 1.3 [33].

Figs. **(5.35)** to **(5.37)** show RANS predictions of the influence of β on particle diffusivity. As the particle diameter increases, the influence of β on the results increases. For the 57μm particles, the use of $\beta = 1.3$ decreases the dispersion by almost 100% while a smaller effect of β is observed on the dispersion of 5μm particles.

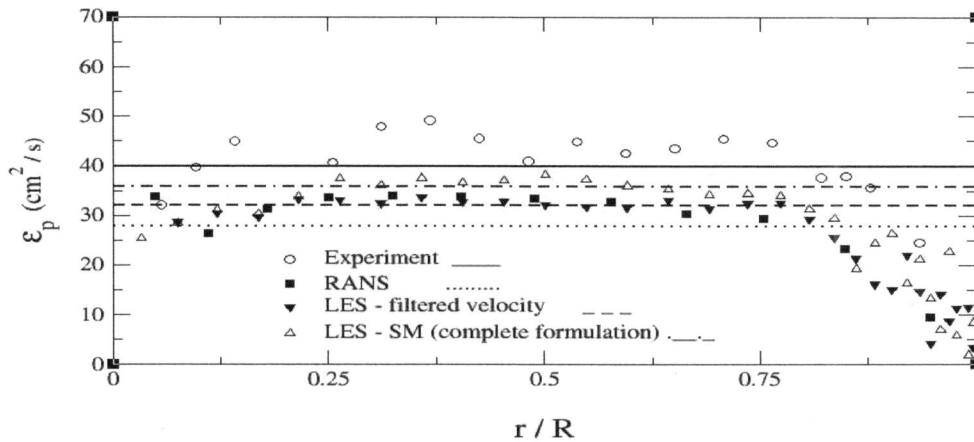

Figure 5.34: Long-time particle dispersion coefficient ϵ_p for $57\mu m$ particles versus the non-dimensionalised pipe radius. Predictions of RANS, LES using only the filtered velocity and LES using the standard formulation of the stochastic model are compared to the experimental observations. Different lines are results computed with Eqn. (5.13). Different shapes of dots are results computed with Eqn. (5.14)

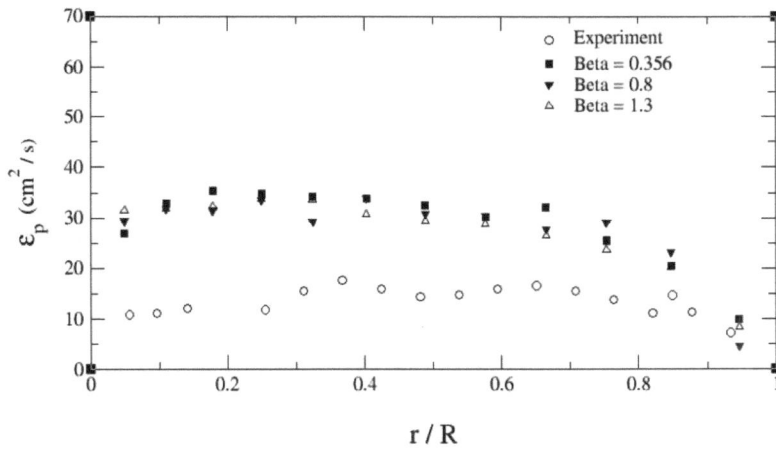

Figure 5.35: RANS predictions of the influence of $\beta = T_L/T_E$ on dispersion coefficient of $5\mu m$ particles.

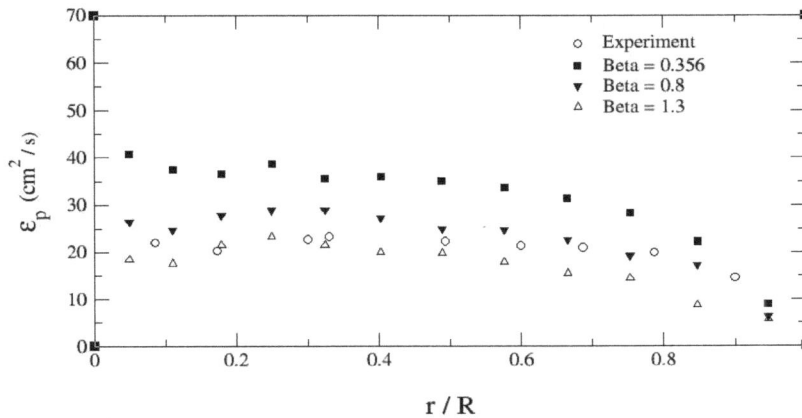

Figure 5.36: RANS predictions of the influence of $\beta = T_L/T_E$ on dispersion coefficient of $37\mu m$ particles.

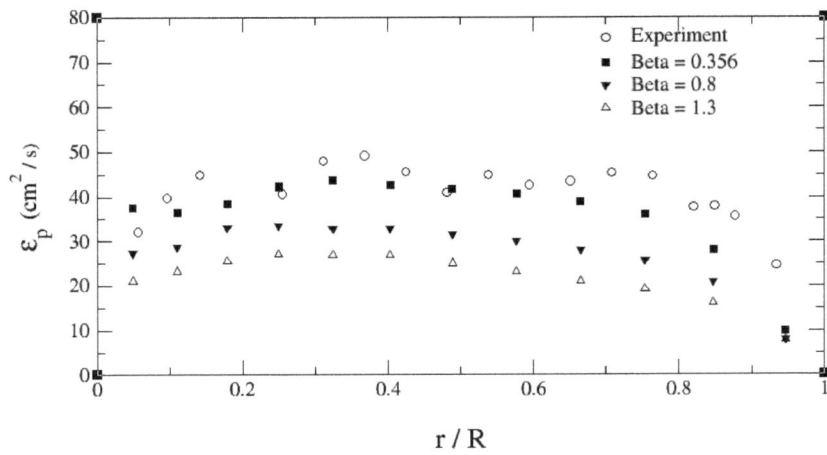

Figure 5.37: RANS predictions of the influence of $\beta = T_L/T_E$ on dispersion coefficient of $57\mu m$ particles.

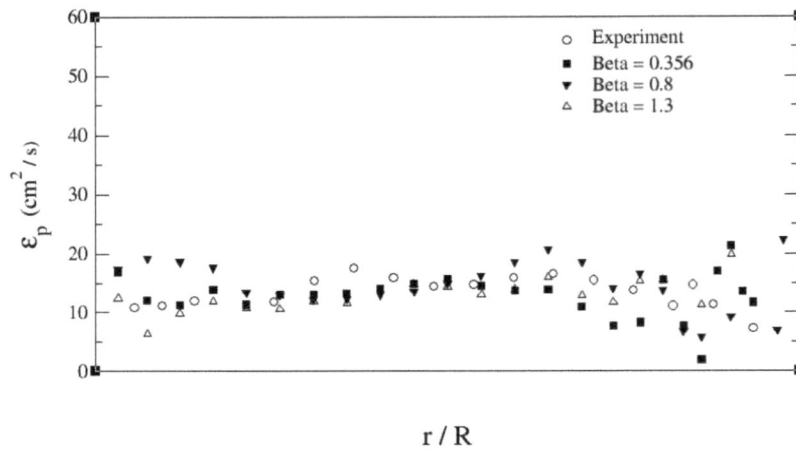

Figure 5.38: LES predictions of the influence of $\beta = T_{L,SGS}/T_{E,SGS}$ on dispersion coefficient of $5\mu m$ particles.

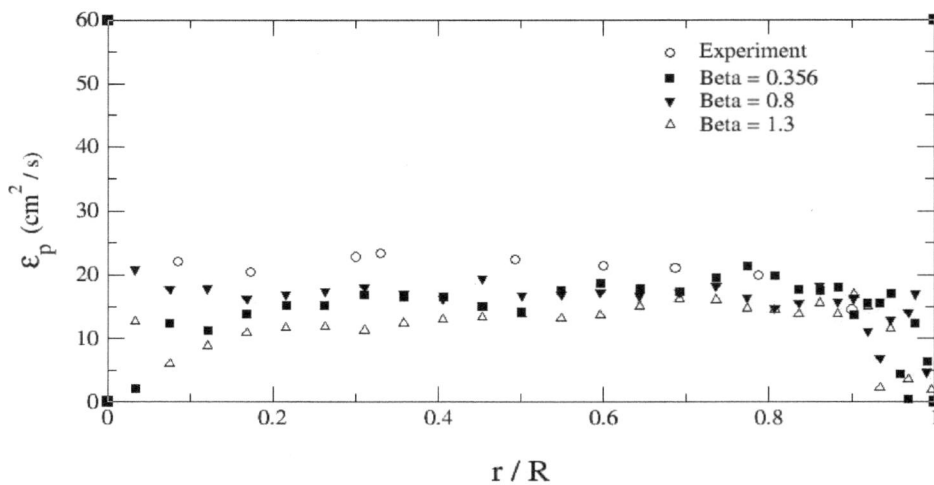

Figure 5.39: LES predictions of the influence of $\beta = T_{L,SGS}/T_{E,SGS}$ on dispersion coefficient of $57\mu m$ particles.

Figs. **(5.38)** to **(5.40)** show LES predictions of the influence of β on particle diffusivity. Unlike RANS, the influence of β on the model predictions is very small though it increases with particle diameter. In fact this tendency is expected in the context of LES since the stochastic model is designed to reconstruct only a small portion of the turbulence fluctuations that are linked to the SGS scales, whereas in the framework of RANS, the model is supposed to rebuild the whole spectrum of turbulence which explains its sensitivity to the values of β.

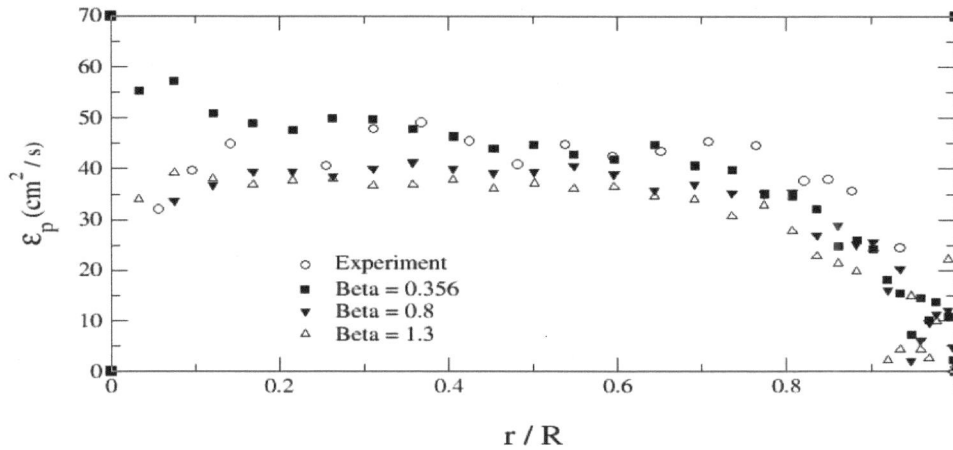

Figure 5.40: LES predictions of the influence of $\beta = T_{L,SGS} / T_{E,SGS}$ on dispersion coefficient of $57\mu m$ particles.

The effect of gravity on particle dispersion in the framework of RANS and LES is also investigated. Presence of gravity results in a finite drift or settling velocity: $u_d = \tau_p . g$, whose magnitude is compared to the slip velocity in the direction of the drift. Fig. **(5.41)** shows the ratio of drift velocity to slip velocity as predicted by RANS and LES for different particle diameters. In fact this ratio also represents the relative importance of the gravity force compared to the drag force in the particle equation of motion. RANS calculations show much a higher ratio at the pipe center compared to LES results. This is due to the differences in predicting particle response time as it is shown in Fig. **(5.18)**. For all particles, the ratio predicted by both simulations is always smaller than unity and decreases quickly as particles move away from the pipe center.

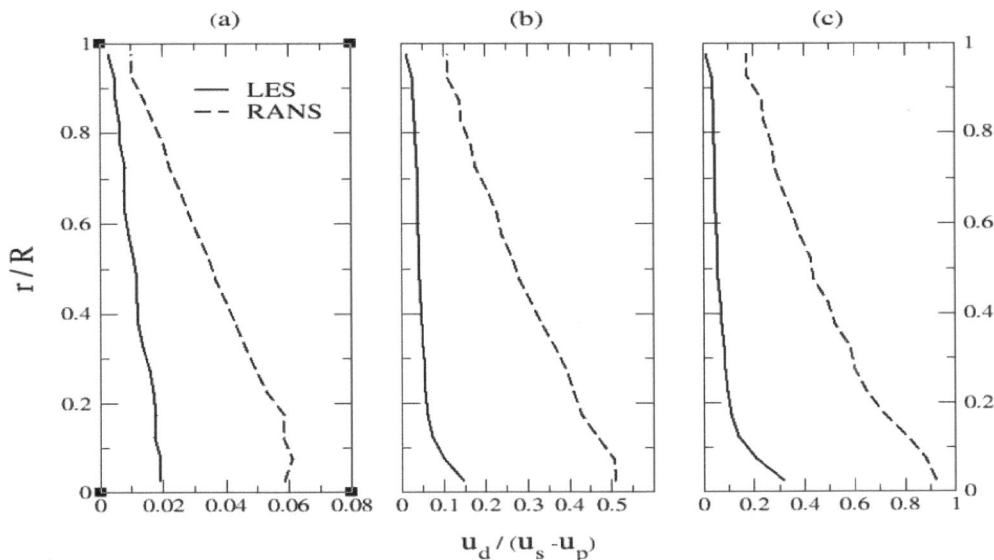

Figure 5.41: RANS and LES predictions of the ratio of drift velocity u_d to streamwise slip velocity $(u_s - u_p)$ versus the non-dimensionalised pipe radius. (a) $5\mu m$ particles, (b) $37\mu m$ particles, (c) $57\mu m$ particles.

Drift parameter is another measurement of the relative importance of the drift velocity to the turbulence level represented by the rms fluctuating velocity. It is defined as the ratio of the drift velocity to the radial rms fluctuating velocity since we are more interested in particle radial dispersion. Fig. **(5.42)** shows RANS and LES computations of the drift parameter for different sizes of particles. Higher values for this parameter are also noticed at the pipe center then it decreases quickly as particles moves towards the wall. It increases with increasing particle diameter but in all cases they are less than unity.

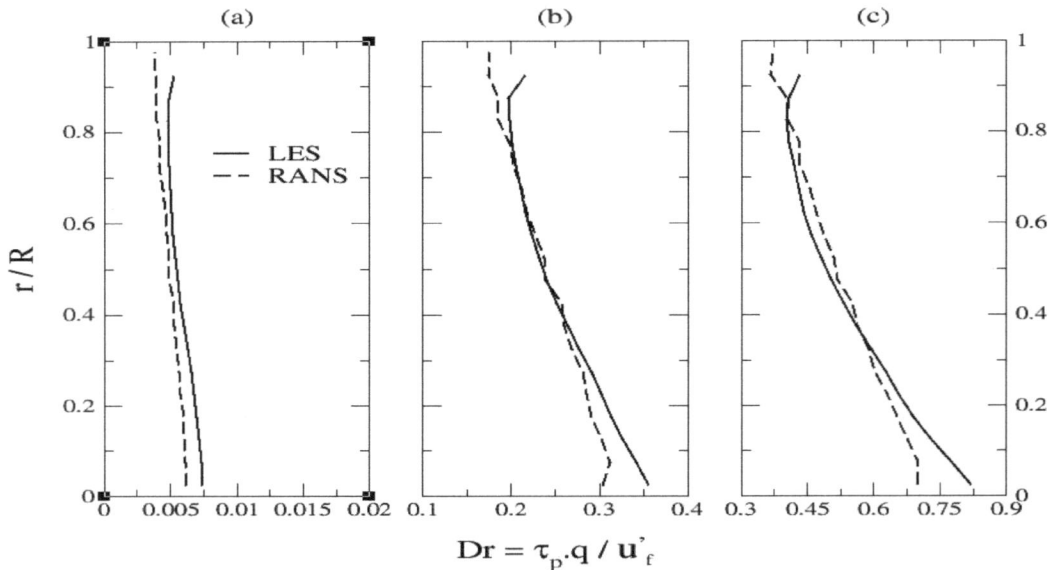

Figure 5.42: RANS and LES predictions of the drift parameter $Dr = \tau_p . q/u'_f$ versus the non-dimensionalised pipe radius. τ_p is the particle response time, q is a body force per unit of mass and u'_f is the fluid rms fluctuating velocity. (a) **5μm** particles, (b) **37μm** particles, (c) **57μm** particles.

Therefore it is concluded that the motion of all the particles is mainly governed by the turbulent flow through the drag force and gravity has little effect on them. Indeed, Figs. **(5.43)** to **(5.45)** show the effect of gravity on particle dispersion in the context of RANS. When the gravity is switched off, only a small increase in particle diffusivity is noticed. Similar conclusion is made concerning LES results as it is depicted in Figs. **(5.46)** to **(5.48)**.

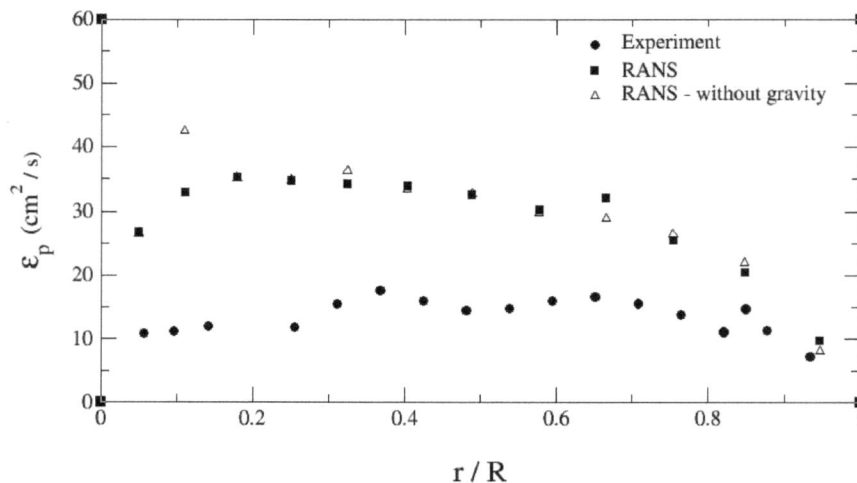

Figure 5.43: Effect of gravity on RANS predictions of the **5μm** dispersion coefficient.

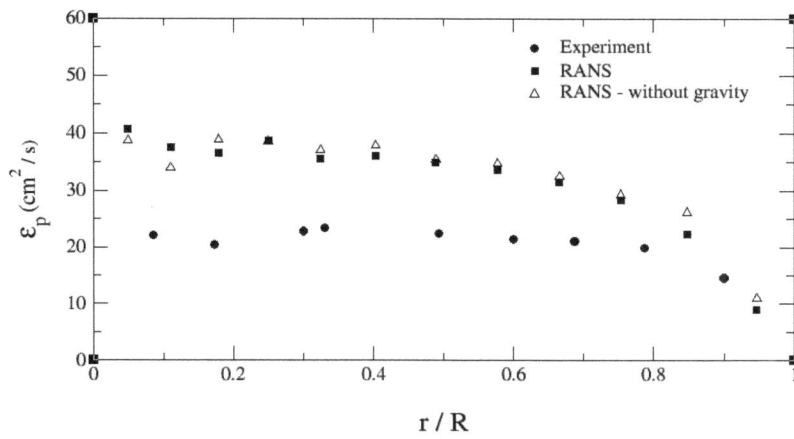

Figure 5.44: Effect of gravity on RANS predictions of the $37\mu m$ dispersion coefficient

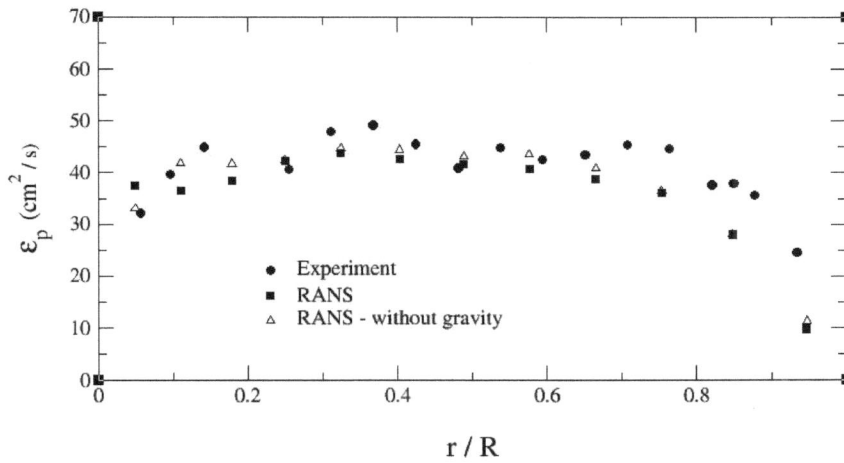

Figure 5.45: Effect of gravity on RANS predictions of the **$57\mu m$** dispersion coefficient.

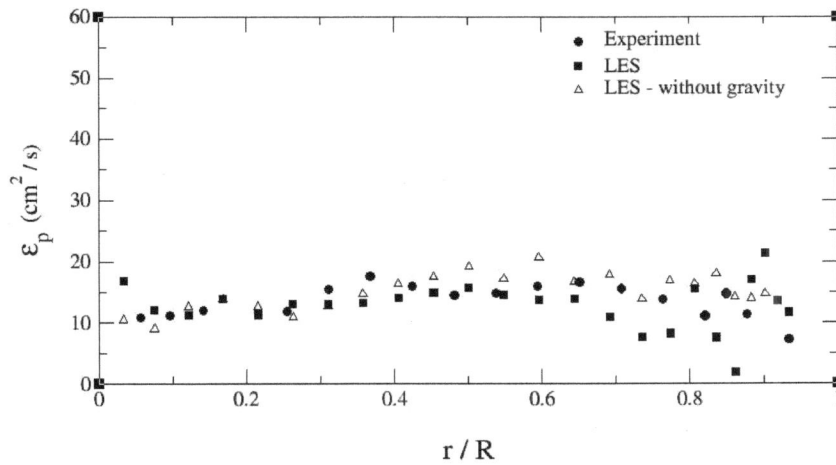

Figure 5.46: Effect of gravity on LES predictions of the **$5\ \mu m$** dispersion coefficient.

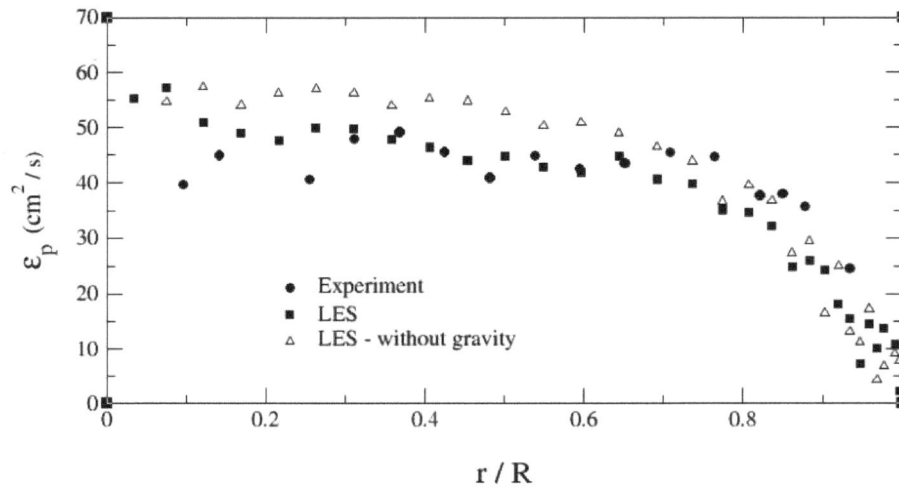

Figure 5.47: Effect of gravity on LES predictions of the **57 μm** dispersion coefficient.

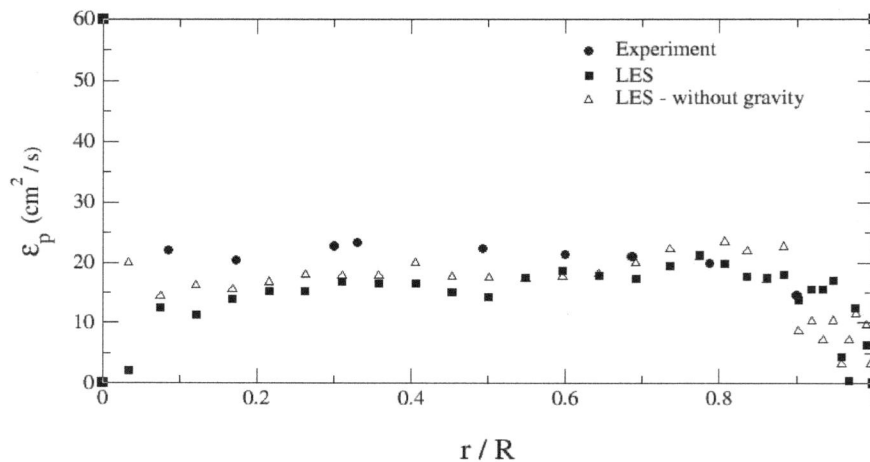

Figure 5.48: Effect of gravity on LES predictions of the **57 μm** dispersion coefficient.

It is postulated that solid particles can sense only turbulent fluctuations they can see with a time scale larger than their response time τ_p. In Figs. **(5.27)** and **(5.28)**, we have shown that for fluctuations occurring at the SGS scales, 5 μm particles do respond to a big portion of SGS fluctuations while 37 μm react only to a smaller portion. 57 μm particles do not sense the SGS scales because their response time scale is larger than the SGS time scale almost everywhere in the flow domain. Similar reasoning is made to examine the influence on particle dispersion of the whole spectrum of turbulent fluctuations that are present in the flow with different time scales. This is can be investigated by computing the time scale seen by particles taking into account both inertia and cross trajectories effects. In Fig. **(5.49)**, RANS and LES predictions of turbulence time scales as seen by solid particles are shown. Both Numerical simulations demonstrate that the inertia and cross trajectory effects increase as the particle diameter increases. LES results predict a more pronounced increase compared to the RANS calculations. This seems to be more logical since both inertia and CT effects are closely linked to particle diameter. Fig. **(5.50)** shows Stokes number of solid particles based on the integral time scale (called also inertia parameter). It is concluded that, like 5 μm and 37 μm particles, 57 μm particles do respond to a portion, though smaller, of the turbulence fluctuations with time scales bigger than those of the SGS scales Numerical simulations are also performed to predict the diffusivity of fluid particles.

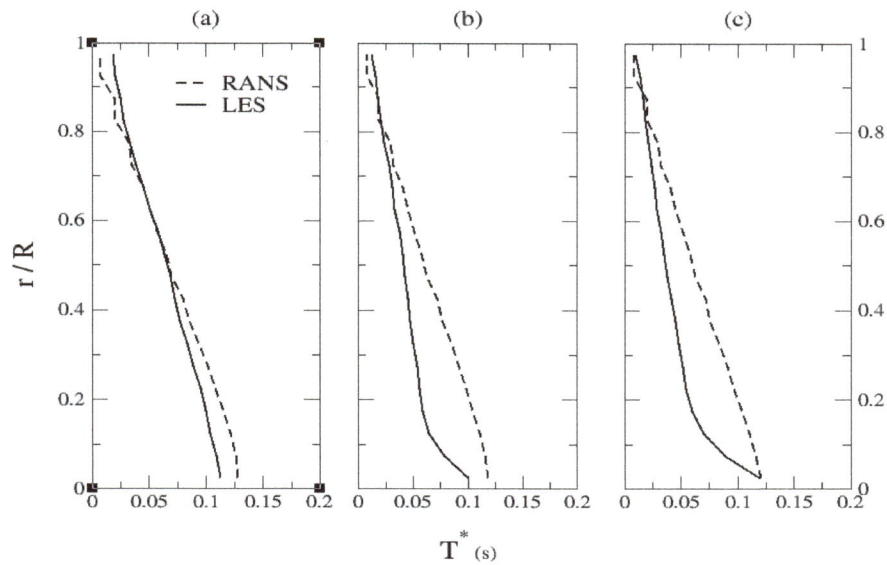

Figure 5.49: Lagrangian time scale T^* with which particles see the total turbulence versus the non-dimensionalised pipe radius. Inertia and cross trajectory effects (CT) are included using Eqns (4.7) and (4.9). (a) **5μm** particles, (b) **37μm** particles, (c) **57μm** particles.

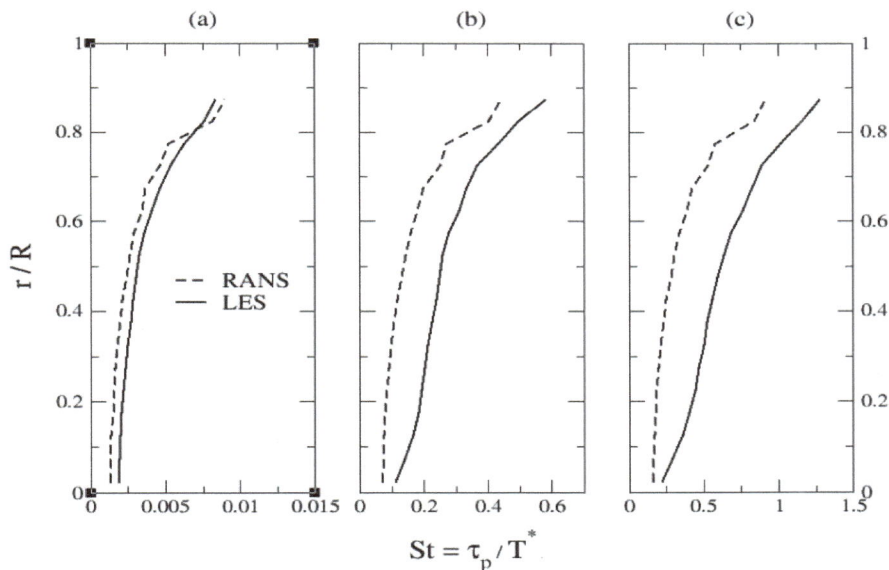

Figure 5.50: Particle Stokes number based on Lagrangian time scale T^* versus the non-dimensionalized pipe radius (a) **5μm** particles, (b) **37μm** particles, (c) **57μm** particles.

Numerical results derived from RANS, LES and LES with subfilter motion included are compared to an empirical estimation of fluid particle diffusivity. Vames and Hanratty [34] stated that the normalized fluid particle diffusivity (with the kinematic viscosity) for one direction divided by the friction Reynolds number should be a constant equal to 0.037; *i.e.* $\epsilon_f / \nu / Re_\tau = 0.037$. This empirical result was derived from several experimental works in pipe flows [21]. It is valid only for tube Reynolds numbers (based on mean velocity) ranging from 10^4 and 10^5. Based on this empirical formula, the fluid particle diffusivity for the present case is computed to be $\epsilon_f \approx 14 \, cm^2/s$. Fig. **(5.51)** shows predictions of fluid element diffusivity as predicted by RANS. As it is expected, they overestimate the fluid particle diffusivity as it was the case for the dispersion coefficients of 5μm particles. In absence of subfilter motion modeling, LES results underestimate the fluid particle diffusivity as it is depicted in Fig. **(5.52)**. Very good agreement with the empirical value of fluid particle diffusivity is obtained when the proposed subfilter stochastic model is used as it is witnessed in Fig. **(5.53)**.

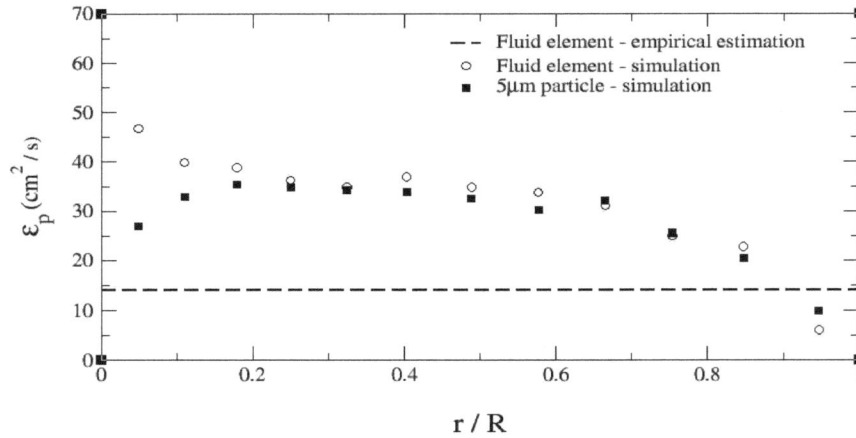

Figure 5.51: RANS predictions of a fluid particle diffusivity versus the non-dimensionalized pipe radius. Empirical estimation of fluid particle diffusivity according to Vames and Hanratty [154] and RANS predictions of dispersion coefficient of **5μm** particle are presented for comparison.

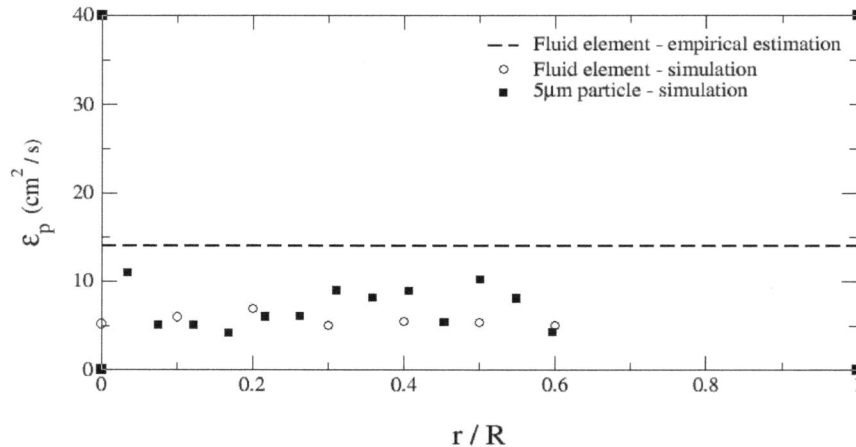

Figure 5.52: LES predictions of a fluid particle diffusivity versus the non-dimensionalized pipe radius. Empirical estimation of fluid particle diffusivity according to Vames and Hanratty [154] and LES predictions of dispersion coefficient of **5 μm** solid particle are presented for comparison. Only the filtered velocity is used to track the fluid particle and the **5 μm** solid particle.

Particle diffusivity is also computed for other particles with larger diameter namely $105\mu m$, $150\mu m$ and $200\mu m$ particles. Fig. **(5.54)** shows an average estimation of solid particle diffusivity non-dimensionalized by fluid particle diffusivity($\epsilon_f \approx 14 \, cm^2/s$) for all particle sizes tracked in the present numerical simulations. Discrepancies between RANS and LES are noticed for particles with small Stokes number ($St < 0.5$) and this is due to the inaccuracy of RANS results for this class of solid particles. The evolution of the normalized particle diffusivity as function of Stokes number depicted in Fig. **(5.54)** is similar to the trend obtained by Tang *et al.* [6] for particle dispersion in spatially developing plane mixing layer. At very small Stokes numbers, solid particles behave like fluid particles and then its diffusivity increases with Stokes number until it reaches a peak near $St = 1$. It keeps this maximum value until around $St = 3$ (10 for the plane mixing layer) and then it decreases as the Stokes number increases. It is important to mention that the results for the heaviest particles ($d_p > 100\mu$) have to be considered with some caution. The use of MR equation to track these inertial particles becomes problematic since their diameters approach the Kolmogorov length scales in particular near the wall (see Fig. **(5.15)** for an estimation of Kolmogorov length scale).

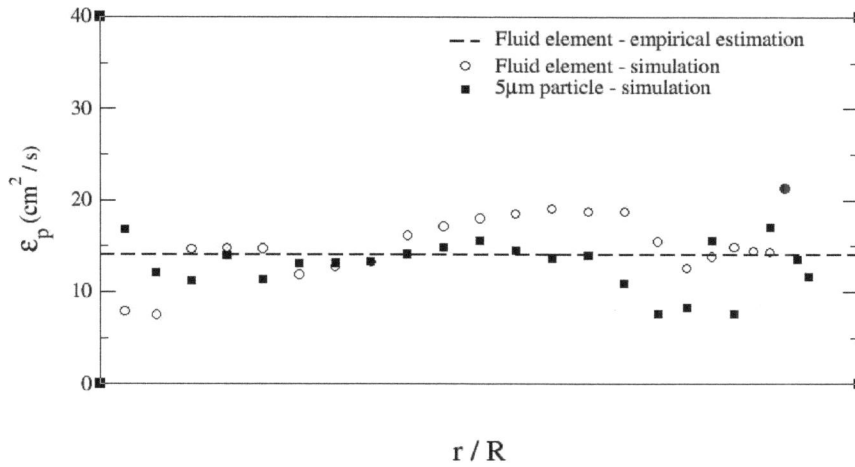

Figure 5.53: LES predictions of a fluid particle diffusivity versus the non dimensionalized pipe radius. Empirical estimation of fluid particle diffusivity according to Vames and Hanratty [154] and LES predictions of dispersion coefficient of 5 *μm* solid particle are presented for comparison. The Standard formulation for the stochastic model is used to track the fluid particle and the 5 *μm* solid particle.

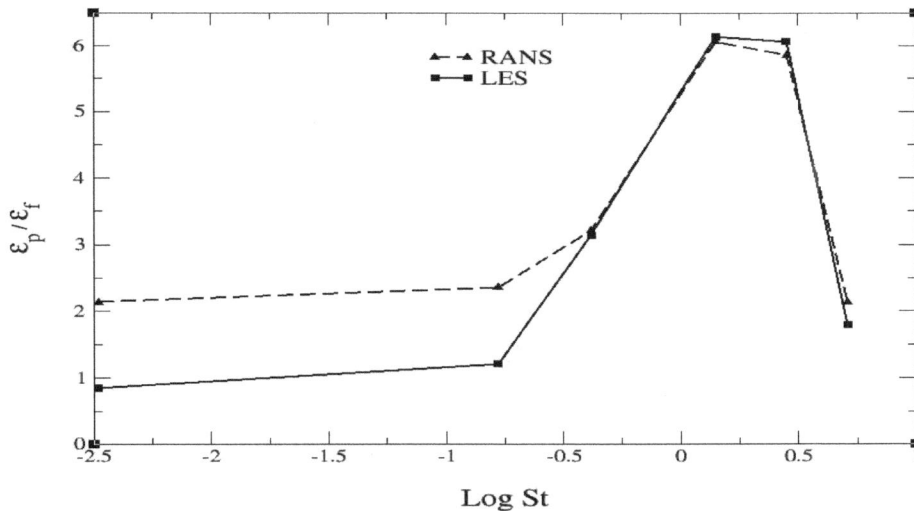

Figure 5.54: RANS and LES predictions of particle dispersion coefficient non dimensionalized by fluid particle diffusivity versus Stokes number in logarithmic scale. $St = \tau_p/T^*$

REFERENCES

[1]   Arnason G. Measurement of particle dispersion in turbulent pipe flow Ph.D. dissertation, Washington State University, Pullman 1982, 183pp.

[2]   Arnason G, Stock DE.Dispersion of particles in turbulent pipe flow. American Society of Mechanical Engineers. Fluid Engin Divi 1983; 10; 25-29.

[3]   Calabrese RV, Middleman S. The dispersion of discrete particles in a turbulent fluid field. AIChE J 1979; 25(6); 1025-1035

[4]   Jones BG. An experimental study of the motion of small particles in a turbulent fluid field using digital techniques for statistical data processing. Ph.D. thesis University of Illinois-Urbana 1966.

[5]   Crowe CT, Gore RA, Troutt TR. Particle dispersion by coherent structures in free shear flows. Part Sci Tech J 1985; 3; 149-158.

[6]   Tang L, Wen F, Yang Y, Crowe CT. Self-organizing particle dispersion mechanism in free shear flows. Phys Fluids 1992; A4(2); 244-51.

[7]   Nir A, Pismen LM. The effect of a steady drift on the dispersion of a particle in turbulent flow. J Fluid Mech 1979; 94; 369-381.

[8]   Reeks MW. On the dispersion of small particles suspended in an isotropic turbulent field. J Fluid Mech 1977; 83; 529-546.

[9]   Wang LP, Stock DE. Dispersion of heavy particles by turbulent motion. J Atmos Sci 1993; 50(13); 1897-1913.

[10]   Taylor GI. Diffusion by continuous movements. Proc Roy Soc 1921; A 151; 421-478.

[11]   Hinze JO. Turbulent fluid and particle interactions. In Progress in heat and mass transfer New York: Pergamon Press 1972; 6; 433-452.

[12]   Reeks MW. Eulerian direct interaction applied to the statistical motion of particles in a turbulent fluid. J Fluid Mech 1980; 97(3); 569-590.

[13]   Lee SL, Durst F. On the motions of particles in turbulent flows. Nuclear Regulatory Commission NUREG/CR-1554. 1980.

[14]   Tennekes H, Lumley JL. A first course in turbulence. MIT Press 1972.

[15]   Smagorinsky, J. General circulation experiments with the primitive equations. Mon Weather Rev 1963; 91; 99.

[16]   Meneveau C. Statistics of turbulence subgrid scale stresses: necessary conditions and experimental tests. Phys Fluids 1994; 6; 815

[17]   Speziale CG. Turbulence modeling for Time-Dependant RANS and VLES: A Review. AIAA J 1998; 36(2)

[18]   Stern F, Wilson RV, Coleman HW, Peterson EG. Comprehensive approach to verification and validation of CFD simulations - Part 1: Methodology and procedure. ASME J Fluids Engine 2001; 123; 793-802.

[19]   Celik IB, Cehreli ZN, Yavuz I. Index of quality for large-eddy simulations. Proceedings of ASME FEDSM2003-45448. 4th ASME JSME joint Fluids Engineering Conference. Honolulu, Hawaii 2003.

[20]   Laufer J. The structure of turbulence in fully developed pipe flow. NACA Rep. 1174. 1954

[21]   Uijttewaal WSJ, Oliemans RVA. Particle dispersion and deposition in direct numerical and large eddy simulations of vertical pipe flows. Phys Fluids 1996; 8(10); 2590-2604.

[22]   Den Toonder JMJ. Drag reduction by polymer additives in a turbulent pipe flow: laboratory and numerical experiments. PhD thesis, Delft University of Technology 1997.

[23]   Zagarola MV. Mean-flow scaling in turbulent pipe flow PhD thesis, Princeton University 1996.

[24]   Laurence D. Length-scales and correlations for RANS-LES coupling Internal Report D3.3-15 for DESIDER project. University of Manchester 2005.

[25]   Sommerfeld M. Modeling of particle-wall collisions in confined gas-particle flows. Int J Multiphase Flow 1992; 18; 905-926.

[26]   Milojevic D, Borner Th. Durst F. Prediction of turbulent gas-particle flows measured in a plane confined jet. PARTEC, Reprints 1st World Congress Particle Technology, Part IV 1986; 485-505 pp.

[27]   Snyder WH, Lumley JL. Some measurements of particle velocity autocorrelation function in turbulent flow. J Fluid Mech 1971 48; 41-71.

[28]   Archambeau F, Mehitoua N, Sakiz M. Code_Saturne: A finite volume code for the computation of turbulent incompressible flows - industrial applications. Int J Finite Vol 2004; 1; 1-62

[29]   Maxey MR, Riley JJ. Equation of motion for a small rigid sphere in a nonuniform flow. Phys. Fluids 1983; 26(4); 883-9.

[30]   Hinze JO. Turbulence. New York: McGraw-Hill 1975.

[31]   Arnason G, Stock DE. New method to measure particle turbulent dispersion using Laser Doppler Anemometer. Experiments in Fluids 1984; 2(2); 89-93.

[32]   Sato Y, Yamamoto K. Lagrangian measurement of fluid-particle motion in an isotropic turbulent field. J Fluid Mech 1987; 175; 183.

[33]   Riley JJ. PhD dissertation, The Johns Hopkins University, Baltimore. Maryland 1971.

[34]   Vames JS, Hanratty TJ. Turbulent dispersion of droplets for air flow in a pipe. Exp Fluids 1988; 6; 94

CHAPTER 6

Large Eddy Simulation of Liquid Particle Deposition in a Turbulent 90° Bend Flow

Abstract: LES was performed to study Aerosol deposition in a turbulent 90° bend flow with tubular cross-section. Numerical predictions were compared to the experimental observations of Pui *et al.* and the DNS-like work of Breuer *et al.* Due to the complexity of the turbulent flow in curved pipe characterized by curved streamlines and zones of recirculation and the lack of comparison studies of the same flow parameters, a good deal of care has been taken to ensure that the carrier phase is accurately simulated. Every effort was made to adapt the mesh to the dynamical features of the flow and boundary conditions were set such that the inlet and outlet conditions would not influence the turbulent flow in the bend.

The numerical predictions of the secondary flow and streamlines in the symmetry plan and in cross sections at different angle of deflection showed a good agreement with the DNS-like work of Breuer *et al. A-posteriori* estimation of the filtered-out kinetic energy demonstrated that the present LES is adequate according to the LES index of quality developed by Celik *et al.*

For the dispersed phase, a stochastic model that accounts for inertial particle transport by SGS motion was used. It was anticipated that such modeling should be crucial owing to the very small-Stokes-number particles tracked. An estimation of the time scale of the SGS fluctuations that are discarded by the filtering operation in LES showed clearly that particles with Stokes number smaller than 0.25 do sense the SGS turbulent fluctuations.

Numerical results concerning the deposition efficiency of inertial particles with Stokes number that range between 0.005 and 1.5 demonstrate the ability of the stochastic modeling to reproduce with good accuracy the SGS effects on small-Stokes-number particles. As it was expected the use of the filtered velocity field only to track particles with Stokes number smaller than 0.3 has proven inaccurate. The complete formulation of the stochastic model showed its superiority compared to the standard formulation. The latter was showed to produce an incorrect level of SGS turbulence.

It was shown clearly that the deposition efficiency of small inertial particles can be predicted with a very good accuracy in the framework of LES using a coarse numerical description. To achieve that, the effect the SGS motion has on inertial particle transport needs to be taken into account. The Langevin-type stochastic diffusion process has proven very adequate in this regard.

Keywords: Bend flows; Large eddy simulation; particle-laden flow; turbulence; Lagrangian description; aerosols; stochastic process; Dean number; deposition; helical vortices.

6.1. INTRODUCTION

In this chapter, liquid particle deposition in a **90°** circular bend is numerically investigated using the stochastic model. Numerical predictions are compared to the experimental observations of Pui *et al.* [1]. In this experimental work, the deposition efficiency of liquid particle in tube bends of circular cross section was measured for flow Reynolds number of 10,000 (based on the bend diameter and mean flow velocity). Fig. **(6.1)** represents an external sketch of the circular bend. The experimental data are summarized in Table **(6.1)**. The experiment was performed using mono-disperse aerosols generated by a vibrating orifice aerosol generator. Depending on the Stokes number of the particles, $0.001 \leq St \leq 1.5$, and the releasing locations at the entrance of the bend, the particles will either deposit on the wall or penetrate and exit the bend. Full description of the test facility, working parameters and data processing are given by Pui *et al.* [1].

The main objective of the experimental work was the determination of the deposition efficiency (fraction of particles deposited in the bend) of liquid particles of different sizes. This deposition takes place in a geometrical and dynamical setting that can be considered as an idealized representation of human respiratory tracts in particular the mouth-throat configuration. Indeed, studies of inertial deposition in a bend have been motivated by interest in calculating the deposition of inhaled particles in human airways. The aim is to help provide more effective treatment of lung diseases, better protection against toxic airborne pollutants, and improvement in routes of systemic drug administration.

Table 6.1: Data of Pui *et al.* Experimental Work [114].

Bend diameter	$D = 0.02m$
Radius of curvature of the bend	$R_b = 0.056m$
Curvature ratio	$R_0 = 5.6$
Centerline inlet velocity	$U \approx 10m/s$
Reynolds number	$Re \approx 10,000$
Dean number	$De \approx 4,225$
Stokes number	$0.001 \leq St \leq 1.5$

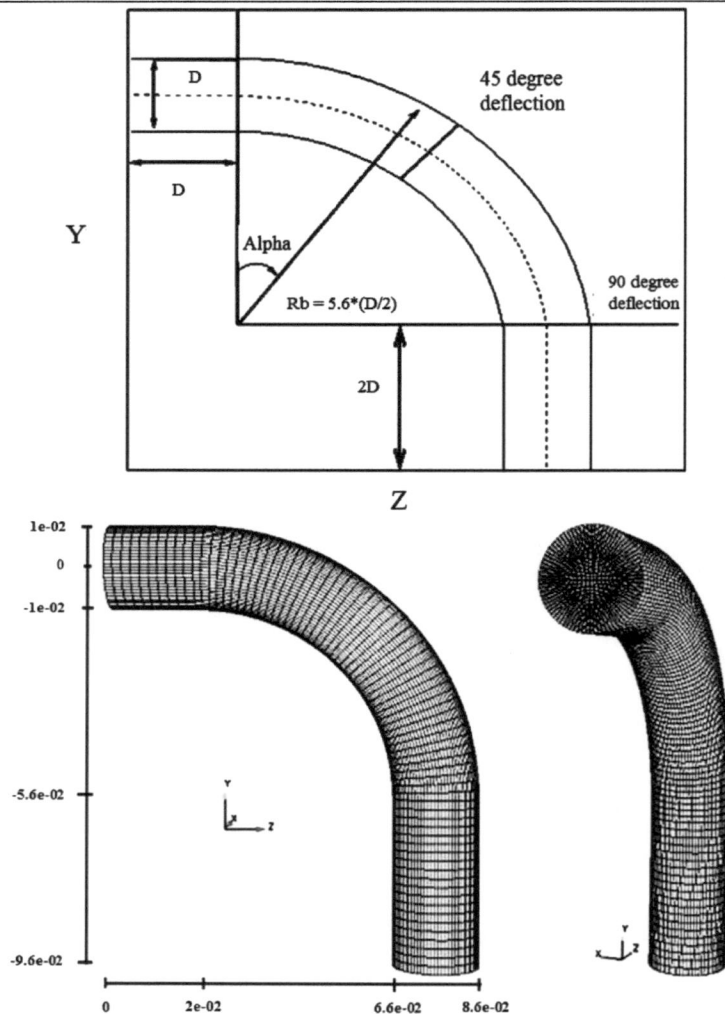

Figure 6.1: External sketch of the circular bend. D is the bend diameter. R_b is the bend radius of curvature. *Alpha* is the angle of deflection.

Other applications consist of systems for sampling aerosol particles from atmosphere or industrial process streams that commonly occur in bends of circular tubes. Significant loss of particles can take place in a bend as a result of inertial deposition. To obtain accurate data, it is important to correct for the losses of particles in bends as well as other parts of a sampling system. This is a major concern for High-Tech industries such as semiconductor manufacturing. For the oil and gas industries, predicting inertial particle deposition and the accompanying erosion phenomena is crucial to avoid extremely expensive component repair, replacement and failure and as a consequence, expensive system shutdown.

The deposition efficiency is defined as the fraction of particles deposited in the bend from the total number of particles that entered the bend,

$$\rho_p = \frac{N_p^{bend}}{N_p^{bend} N_p^{after\ bend}}. \tag{6.1}$$

From dimensional considerations, it can be shown that the deposition efficiency of a bend for particles of finite sizes depend on the following five dimensionless groups [2]:

1. Strokes number : $St = C_n \rho_p d_p^2 U_0 / 18 \mu R$,

2. particle Reynolds number : $Re_p = d_p U_0 / v$,

3. interception parameter : $I = d_p \backslash 2R$,

4. flow Reynolds number : $Re = 2R U_0 / v$,

5. curvature ratio : $R_0 = R_b / R$.

In the above definitions $\rho_p, d_p, U_0, \mu, v, C_n, R, R_b, N_p^{bend}, N_p^{after\ bend}$ are particle density, particle diameter, mean axial velocity, fluid dynamic viscosity, fluid kinematic viscosity, Cunningham slip correction factor, tube radius, radius of curvature of the bend, number of particles that deposit in the bend and number of particle that exit the bend respectively.

Among these five groups the Stokes number and the flow Reynolds number are the most important parameters [1]. The particle Reynolds number can be an important parameter in the ultra-Stokesian regime, *i.e.* $Re_p > 1$. In practical situations, the particle size is several orders of magnitude less than the tube diameter. Thus, the effect of interception parameter is negligible. For $4 \leq R_0 \leq 30$, the effects of curvature ratio are not significant [2]. For turbulent flows, the deposition efficiency does not depend on the flow Reynolds number. For laminar cases, deposition efficiency is sensitive to variation in the flow Reynolds number [3].

As far as the inertial particle-turbulence interaction is concerned, it is well understood that small particles easily follow the large scale turbulent motions. They are therefore most sensitive to the variations in turbulence properties as occurring in the near-wall layer of a tube flow. The motion of large particles is more dominated by the overall turbulence characteristics and is less sensitive to the presence of a near-wall layer. This particle-turbulence interaction predominantly affects the particle deposition in wall-bounded turbulent flows. Therefore, turbulence modeling for the carrier phase and Lagrangian trajectory models for the dispersed phase need to be as accurate as possible in order to accurately predict particle deposition.

The use of RANS turbulence models to predict the continuous phase has proven adequate in flows for which they have been calibrated, mainly high Reynolds number and boundary layer flows. The flow taking place in bends has many characteristics such as curved streamlines and recirculation regions. These dynamical features are known to cause two-equation turbulence models to perform poorly [4]. For the dispersed phase, the Random Walk/Eddy Interaction models are usually used to compute particle trajectories in turbulent flows. These models work well for a narrow range of flows for which they are calibrated such as single particles in homogeneous grid turbulence. However, they are expected to perform poorly for the case in hand since they rely heavily on the turbulence modeling designed for the carrier phase. For the same reason, the stochastic diffusion processes are not expected to perform better though they are mainly developed for non-homogeneous and anisotropic flows.

Overcoming the above shortcomings may be possible by using more advanced turbulence models such as large eddy simulation. In LES, only information about small scales are filtered out and their influence on large scale turbulence is modeled. In principle, this should be sufficient to predict turbulent fluctuations and their effects on the transport of inertial particles. However, in wall-bounded turbulent flows, the filtered-out small scales near the wall often carry a substantial amount of kinetic energy. The spatial resolution in these regions is not usually fine enough to capture them, thus resulting in an excessive filtering-out of kinetic energy. This could be also the case for flows with recirculation zones and secondary motion. Modeling turbulent fluctuations linked to these small scales has proven crucial for small-Stokes-number inertial particle dispersion as it was demonstrated in Chapter 5. This modeling is expected to be equally important for particle deposition.

The Langevin-type stochastic diffusion process that is described in Chapter 4 and used for particle dispersion prediction in a turbulent pipe flow in Chapter 5, is tested in this chapter to predict particle deposition in a 90 bend. Both RANS and LES calculations for the deposition efficiency are compared to the experimental work of Pui *et al.* [1].

6.2 GOVERNING EQUATIONS

6.2.1 Continuous Phase

In finite bends of circular cross-section, the turbulent flow dynamics is complex. Indeed, it is characterized by the existence of recirculating regions and curved streamlines. If the flow in the bend is meant to represent the flow in the human airways such as the throat, the Reynolds number is of the order of a few thousands. So transitional flow that is neither fully laminar nor fully turbulent may occur. This fact has been observed in many experimental results [1].

For curvature ratio R_0 greater than five, the flow field in bends of circular cross section depends only on the Dean number [2]. The Dean number is defined as the flow Reynolds number divided by the square root of the curvature ratio:

$$De = Re/\sqrt{R_0}, \tag{6.2}$$

The curvature ratio is defined as $R_0 = R_b/R$ where R_b is the radius of curvature of the bend and R is the tube radius. The Dean number represents the ratio of the square root of the product of centrifugal forces (U^2/R_b) and inertial forces (U^2/R) to the viscous forces $(\nu U/R^2)$. It plays the role of the Reynolds number of the flow in a curved pipe.

Based on the Dean number, three different flow regimes for the flow in circular bends have been identified. For small Dean numbers $(De \leq 17)$, a pair of counter-rotating helical vortices, placed symmetrically with respect to the plane of symmetry are formed. This is due to the centrifugally induced pressure gradient. It drives the slower moving fluid near the wall inward, while the faster moving fluid in the core is swept outward. The position of the maximum axial velocity moves toward the outer bend.

For intermediate Dean numbers $(17 \leq De \leq 370)$, the flow pattern is the same as before but with the peak velocity closer to the outer bend. Also, a distortion of the secondary streamlines occurs. In this regime, especially as the Dean number gets larger, the structure of the flow is characterized by an inviscid rotational core surrounded by a thin boundary layer. The flow is considered laminar at least for small and intermediate Dean numbers $(De \leq 370)$.

For larger Dean number $(De > 370)$ secondary boundary layers develop on the wall, with fluid entering these boundary layers near the outer bend and leaving near the inner bend. Further increase of the Dean number, and hence the centrifugal force, leads to an increase in axial circumferential velocity and to more fluid being sucked into the secondary boundary layers near the outer bend. The secondary boundary layers adjust by thinning near the outer bend and thickening near the inner bend. Simultaneously, the locations of the maximum axial velocity moves toward the outer bend. The two vortices are skewed by the adjustment of the secondary boundary layers.

These dynamical features that characterize a turbulent flow in a circular bend are expected to be captured by solving the Navier-Stokes equations for all the scales down to Kolmogorov scales. This is the direct numerical simulation and it is an expensive approach. RANS and LES are two practical alternatives that rely on modeling of the whole turbulence for the former and the one related to the small scales for the latter. Both approaches are used to compute the velocity field for the carrier phase of the test case in hand.

The LES formulation for the continuous phase has been developed in Chapter 2 and used to simulate the high Reynolds number turbulent pipe flow in Chapter 5. Previous simulations of fully developed straight pipe flow with the Smagorinsky model and the Van Driest damping function have shown good agreement with experiments as it was shown in Chapter 5. From turbulence modeling we know that models based on eddy viscosity assumptions do not perform well in flows with curved streamlines. This may perhaps leads one to expect that the Smagorinsky subgrid model is not applicable for LES of curved pipe flow. However, Eggels *et al.* [5] showed that it correctly predicts the large scales in a rotating pipe flow for which the eddy viscosity assumption is totally inappropriate. Therefore, it can be applied to curved pipe flow without much adverse effects on the simulation of the general characteristics of the flow.

The Reynolds Averaged Navier-Stokes (RANS) equations are obtained by time-averaging the continuity and Navier-Stokes equations governing viscous incompressible flow,

$$\frac{\partial U_i}{\partial x_i} = 0,$$ (6.3)

$$\frac{\partial U_i}{\partial t} + U_j \frac{\partial U_i}{\partial x_j} = -\frac{1}{\rho_f} \frac{\partial P}{\partial x_i} + \nu \frac{\partial^2 U_i}{\partial x_j \partial x_j} - \frac{\partial <u_i u_j>}{\partial x_j}.$$ (6.4)

The total velocity is decomposed into a sum of its mean and a fluctuation: $u_i = U_i + u_i'$, noting the rules $< U_i >= U_i$ and $< u_i' >= 0$. The pressure field also is divided to a mean P and pressure fluctuation p'. The mean flow equations (6.3) and (6.4) are unclosed because of the extra 6 unknowns that are the components of the Reynolds stress tensor: $< u_i u_j >$. For the second order closure of RANS equations such as Reynolds Stress Model (RSM), transport equations are written for the Reynolds stress components as depicted in equation (6.5):

$$\frac{\partial < u_i u_j >}{\partial t} + U_k \frac{\partial < u_i u_j >}{\partial x_k} = -\frac{1}{\rho_f} (< u_j \frac{\partial p}{\partial x_i} > + < u_i \frac{\partial p}{\partial x_j} > -2\nu < \nabla u_i \nabla u_j >$$

$$+ \nu \nabla^2 < u_i u_j > -< u_j u_k > \frac{\partial U_i}{\partial x_k} -< u_i u_k > \frac{\partial U_j}{\partial x_k}.$$

$$-\frac{\partial < u_k u_i u_j >}{\partial x_k}$$ (6.5)

Redistribution, turbulent transport and dissipation need to be modeled but the important issue is that the production terms can be computed exactly. These production terms represent the effects of rotation and streamline curvature on turbulence production or damping. This is why the Reynolds stress models have a larger potential to represent the turbulent flow features more correctly than the two-equation models. Excellent results have been obtained for some flows in which the $(k - \epsilon)$ type of models are known to perform badly such as flows with curvature and separation from curved surfaces [4]. This is mainly due to the fact that Reynolds stress models are based on the Reynolds stresses transport equations rather than on the crude eddy viscosity assumption on which rely two-equation models. This inherently assumes an instantaneous equilibrium between the Reynolds stress tensor and the mean rate of strain. This assumption does not hold for rapidly changing flow conditions and disequilibrium should come into play [6]. Moreover, Reynolds stresses models account for the normal stress anisotropy unlike the two-equation models that incorrectly represent the turbulence anisotropy through a scalar; the kinetic energy [7]. The different unclosed terms in Equation (6.5) are modeled using some the popular models such as (LRR) of Launder, Reece and Rodi [8] and the one developed by Speziale, Sarkar and Gatski [9].

Another important issue in turbulence modeling for RANS is the numerical treatment of the equations in regions close to walls. Indeed, the near wall formulation determines the accuracy of the wall shear stress. It has an important impact on dispersion and deposition of inertial particles in two-phase turbulent flow. Often, the RSM model is used in conjunction with *scalable wall functions*. They are the only available formulation that allows users to apply arbitrarily fine grids without violating the underlying logarithmic profile assumptions [10].

To eliminate the unknown influence of the inlet and exit conditions on the flow development in the bend, a horizontal inlet section of length D should be mounted to the 90° bend. The outlet is elongated by adding a vertical straight pipe with a length of 2D. The entire geometry is shown in Fig. **(6.1)**. To avoid the use of artificial inflow data which often only partially satisfy all physical requirements, an appropriate inflow boundary conditions for the turbulent flow within the bend is needed. For that purpose, a separate LES prediction of a 5D-length straight pipe with the same cross section, flow Reynolds number and time step size is usually carried out. Periodic conditions are used and the length of the straight pipe (5D) allows all the two-point correlations to vanish midway between the boundaries. It is known that the flow becomes fully developed when it no longer varies in the streamwise direction. When this requirement is achieved, the velocity profile of one cross-section of this periodic LES is used as inflow conditions at the inlet of the pipe mounted in front of the bend. This ensures that the inlet conditions that are prescribed at the inflow boundary of the bend satisfy the Navier-Stokes equations and the level of turbulence intensity is physically correct. If not fully satisfied, this should not affect significantly the flow in the bend. This is due to the fact that in the curved pipe case the flow is strongly mixed by the secondary motion. Thus, the transient from the initial conditions disappear faster than for the case of a straight pipe where the transient from the initial conditions persists for a longer time. For the inflow conditions needed for the RANS calculations, mean velocity field, different components of the Reynolds stress tensor and the dissipation profile are stored. As a result, a fully developed flow profile is obtained prior to the bend for both RANS and LES calculations.

Non-reflecting boundary conditions are usually applied at the outlet of the elongated bend. This ensures that all vortical structures can leave the integration domain without significant disturbances or wave reflections into the inner region. Since the first grid point is placed sufficiently close to the wall, it is possible to apply the no-slip boundary condition at the wall. The pressure is computed assuming a vanishing pressure gradient in the wall-normal direction.

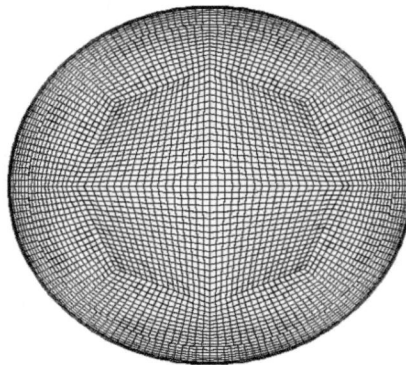

Figure 6.2: Cross-section grid distribution used for LES.

Table 6.2: Non-Conforming Embedded Refinement in the Polar Part of the Unstructured Grid of the Bend.

$Y^+ = u_\tau . y/\nu$	Rad. Direction	Circumf. Direction
$0 < Y^+ < 5$	3 cells	144 cells
$5 < Y^+ < 30$	4 cells	120 cells
$3 -< Y^+ <\approx 180$	10 cells	96 cells

For LES, an unstructured grid consisting of 1,300,000 cells is used with about 4,700 grid points on each cross section. A polar grid is used for the first three layers with non-conforming embedded refinement as shown in Table **(6.2)**. Then the polar grid is made to match an octahedral bloc for the core region of the pipe as it is depicted in Fig. **(6.2)**. The 2D grid is then extruded in the streamwise direction by 240 nodes as it is shown in Fig. **(6.3)**. The first grid point near the pipe wall at which the axial velocity is computed is located at $y^+ \approx 1.3$ grid points placed within the viscous sublayer, the depth of which equals 5 wall units. A non-uniform grid is employed in the normal-to-the-wall direction within the circular part. This is done in order to locate more grid-points in the near-wall region that is characterized by steep gradients and small energy-containing eddies. It is worthy to note that two grid points in the viscous sublayer is in general enough for LES of straight pipe flow. For curved pipe the thickness of the viscous sublayer is non-uniform. Indeed for the fully developed curved pipe, the sublayer thickness at the outside radius is approximately 3.5 wall units while at the inside radius the depth of the viscous layer measures 8 wall units. By having 3 points under 5 wall units everywhere in the domain, the resolution of the viscous sublayer regardless of its location should be adequate especially at the outside of the bend. It is also clear that the resolution at the inside of the bend and in the straight pipes will be increased since more than 2 points are used to resolve the viscous sublayer.

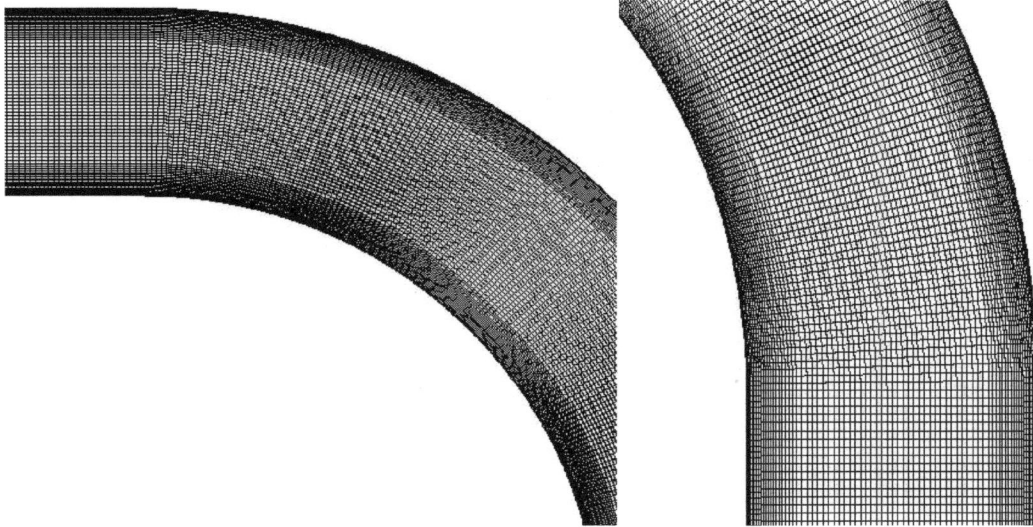

Figure 6.3: Streamwise grid distribution used for LES.

Care should be taken when using non-conforming embedded refinement to locate cells where more resolution is needed. Indeed, the near wall region is characterized by steep gradients and very small energy-containing eddies that should be well captured. These near-wall coherent structures contain most of the turbulence and are responsible for the correct distribution of the turbulent energy from the streamwise into the other directions. Moreover, the near-wall turbulence has a significant impact on the deposition of inertial particles and therefore it should be properly resolved. As depicted in Figs. **(6.4)** to **(6.8)**, the different non-conforming layers are designed such that the aspect ratios between the different directions for all the cells do not exceed 5. This is done to avoid numerical instabilities that arise when using very flat cells in the LES context. Thus the use of non-conforming layers should be efficient since it allows a cell distribution that responds to the requirement of the flow dynamics without introducing further numerical errors.

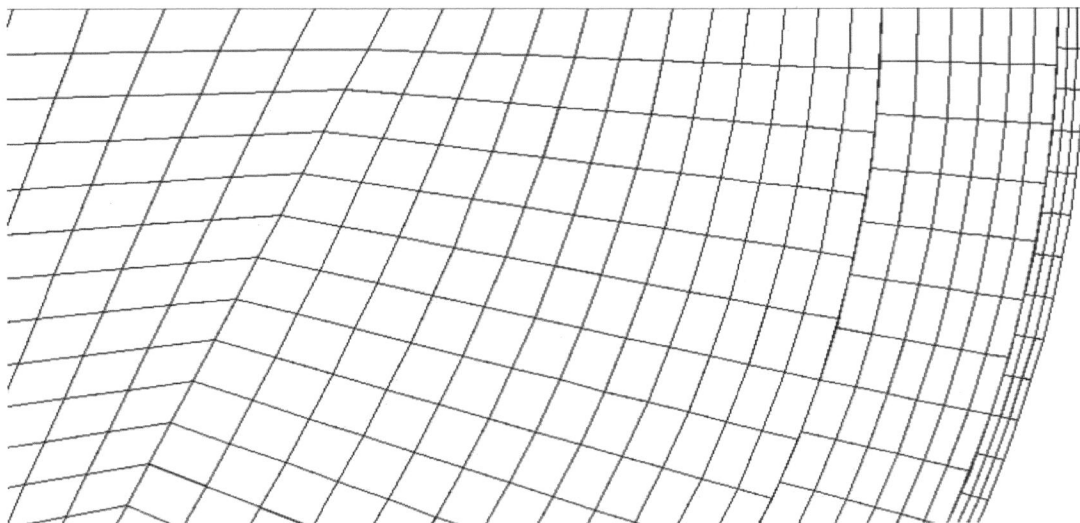

Figure 6.4: Non-conforming layers in the radial direction.

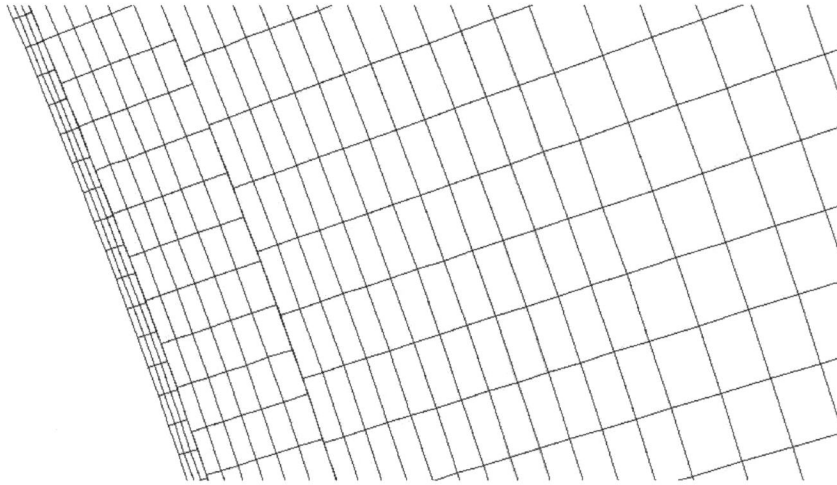

Figure 6.5: Non-conforming layers in the streamwise direction.

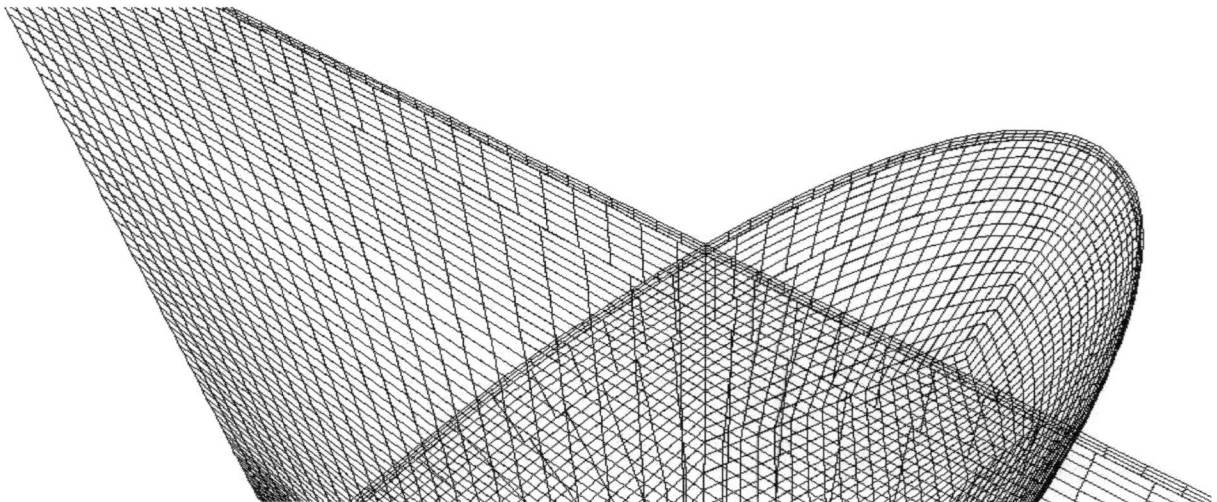

Figure 6.6: Use of non-conforming embedded refinement for LES.

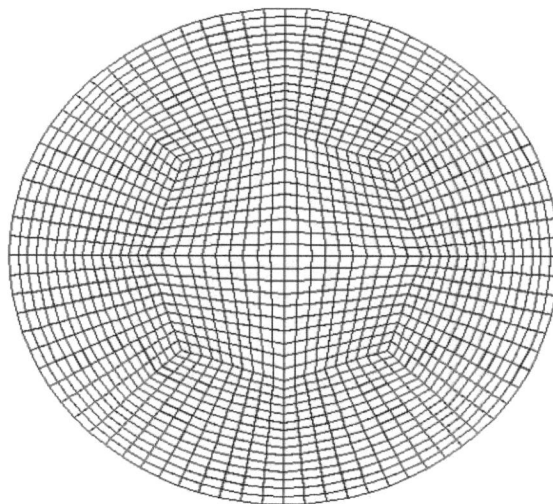

Figure 6.7: Cross-section grid distribution used for RANS.

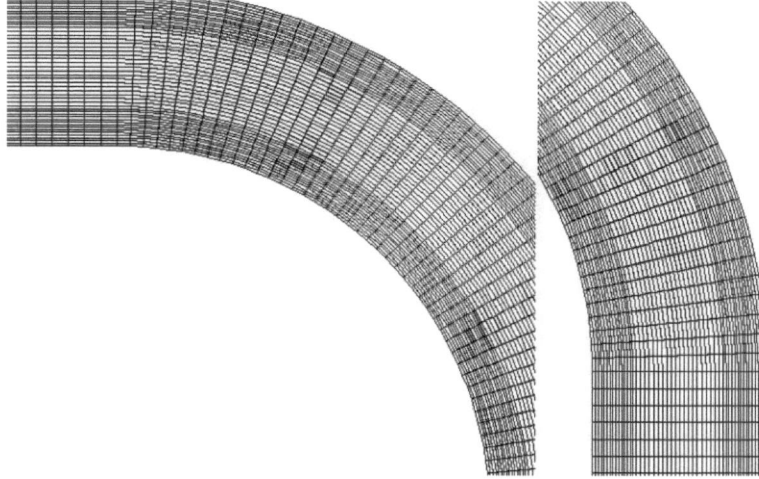

Figure 6.8: Streamwise grid distribution used for RANS.

The LES calculations performed by Breuer *et al.* [3] for the same case used a structured grid that consisted of 2,280,000 cells with 8,900 grid points in each cross section. According to Breuer *et al.* [3] the resolution is found sufficient to resolve all relevant scales at the moderate Reynolds number considered in the present study. No evidence was given to underpin their claim such as an estimation of the amount of the filtered-out kinetic energy. We refer herein to this work as the DNS-like work of Breuer *et al.* [3]

For RANS, the cross section is discretized with about 1,600 grid points. The grid used here contains approximately 120,000 nodes. Scalable wall functions are used with the center of the first control volume at $y^+ \approx 11$. Grid convergence was determined by comparing the base grid with a low-resolution grid (60,000 grid points) and a high-resolution (160,000 grid points).

6.2.2 Dispersed Phase

For Lagrangian particle tracking used in conjunction with RANS calculation of the carrier phase, the use of the stochastic approach to predict the fluid velocity seen by the inertial particle along their trajectories is based on the mean velocity fluid (instead of the filtered velocity field for LES) and the turbulent kinetic energy (Reynolds stress tensor) and its dissipation rate provided by RANS calculus of the carrier phase (instead of the residual kinetic energy and its dissipation rate for LES).

The standard and complete formulations of the Langevin-type stochastic equations are discussed in the framework of LES in Chapter 4. Also, they are used to predict particle dispersion in a turbulent pipe flow in Chapter 5. For RANS calculations, the standard and complete formulations are the following:

The standard Formulation

$$du_{s,i} = \left(-\frac{1}{\rho_f}\frac{\partial P}{\partial x_i} + \frac{1}{Re}\frac{\partial^2 U_i}{\partial x_j \partial x_j}\right).dt - \left(\frac{u_{s,i}-U_i}{T^*}\right).dt + \sqrt{C_0 <\epsilon>}\, dW_i. \qquad (6.6)$$

The complete formulation

$$du_{s,i} = \left(-\frac{1}{\rho_f}\frac{\partial P}{\partial x_i} + \frac{1}{Re}\frac{\partial^2 U_i}{\partial x_j \partial x_j}\right).dt - \left(<u_{p,j}> -U_j\right)\frac{\partial U_i}{\partial x_j}.dt - \left(\frac{u_{s,i}-U_i}{T^*}\right).dt + \sqrt{C_0^* <\epsilon>}\, dW_i. \quad (6.7)$$

These two formulations are used to stochastically construct the fluid velocity seen by the inertial particles along their trajectories. They are solved simultaneously with the particle equations of motion:

$$dx_{p,i} = u_{p,i}.dt,$$

$$du_{p,i} = \frac{u_{s,i} - u_{p,i}}{\tau_p}.dt + g.dt, \tag{6.8}$$

$$\tau_p = C_n \frac{\rho_p}{\rho_f} \frac{4d_p}{3C_D|u_s - u_p|},$$

$$C_D \begin{cases} \frac{24}{Re_p}\left(1 + 0.15Re_p^{0.687}\right) \ if \ Re_p < 1000 \\ 0.44 \ if \ Re_p > 1000. \end{cases} \tag{6.9}$$

C_n is Cunningham slip correction factor. It is considered herein to account for the Brownian diffusion for particles smaller than 1 micron. Indeed, very small particles in a gas stream deflect slightly when gas molecules strike them. Transfer of kinetic energy from the rapidly moving gas molecules to the small particles causes this deflection that is called Brownian diffusion.

For Lagrangian two-phase RANS, the stochastic approach is designed to reproduce the whole turbulence field washed out by the time averaging. For the case in hand, this turbulence field is anisotropic. Therefore the directional dependence of the Lagrangian time scale T_L has to be taken into account. Oesterle and Zaichik [11] developed a directional estimation of the Lagrangian time scale T_L based on the Taylor's theory on fluid particle diffusion [12]. They identify the eddy viscosity ν_t with the long time turbulent diffusivity leading to the following directional estimation of T_L :

$$T_{L,1} = \nu_t / < u_1 u_1 >. \tag{6.10}$$

$$T_{L,2} = \nu_t / < u_2 u_2 >. \tag{6.11}$$

$$T_{L,3} = T_{L,2} < u_3 u_3 > / < u_2 u_2 >. \tag{6.12}$$

Here direction 3 is the direction parallel to the mean drift and directions 1 and 2 are directions perpendicular to the mean drift direction.

As explained in Chapter 4, the inertia effect is accounted for using Wang and Stock formula [13],

$$T_i^* = \frac{T_{L,i}}{\beta}\left(1 - (1 - \beta)(1 + St)^{-0.4(1+0.01St)}\right), \tag{6.13}$$

where β is the ratio between the Lagrangian and the Eulerian time scales taken equal to 0.356 [13].

As far as the cross-trajectory and continuity effects are concerned, Csanady formulae (1963) are used:

$$T_1^* = \frac{T_1^*}{\sqrt{1 + 4\beta^2| < u_r > |^2(2k/3)}},$$

$$T_2^* = \frac{T_2^*}{\sqrt{1 + 4\beta^2| < u_r > |^2(2k/3)}},$$

$$T_3^* = \frac{T_3^*}{\sqrt{1 + \beta^2| < u_r > |^2(2k/3)}},$$

Here $< u_r >$ is the mean slip velocity between fluid and inertial particles.

The diffusion constant is evaluated as it is shown in Chapter 4. for LES calculation. The only difference is that the kinetic energy used in the framework of RANS is the whole turbulent kinetic energy computed using the diagonal terms of the Reynolds stress tensor:

$$k = \frac{1}{2}(<u_1u_1> + <u_2u_2> + <u_3u_3>), \tag{6.14}$$

$$C_0^* = C_0 b_i \tilde{k}/k + \frac{2}{3}(b_i \tilde{k}/k - 1), \tag{6.15}$$

where $\tilde{k}$ is a modified kinetic energy,

$$\tilde{k} = \frac{3}{2} \frac{\sum_{i=1}^{3} b_i <u_i'^2}{\sum_{i=1}^{3} b_i} \tag{6.16}$$

Here u_i' is the fluid fluctuating velocity. $b_3 = \sqrt{1 + \beta^2 |<u_r>|^2/(2k/3)}$ for the mean drift direction. $b_{1,2} = \sqrt{1 + 4\beta^2 |<u_r>|^2/(2k/3)}$ for the perpendicular directions 1 and 2.

To ensure particle statistics of high quality, the number of released particle was set to a large value, 250,000 particles for each class of particle size. They are randomly distributed over the cross section at the inlet and released at the first time step. Then they are tracked throughout the flow field until they exit the bend or deposit at the wall.

It is considered that particles adhere to the surface upon contact. Numerically, contact is supposed to occur when the particle radius is larger or equal to the normal distance between the surface and the particle center. No bounce from the surface is considered since the surface is considered wet as it is the case for the mouth-throat region. Though a large number of particles are released and tracked within the flow, neither two-way coupling nor particle collision is considered. This was the case of the experimental work of Pui *et al.* [1] since they used a very dilute suspension. The number of particles tracked for these numerical simulations is set reasonably large in order to reduce the statistical noise on the deposition results.

6.3 RESULTS AND DISCUSSION

6.3.1 Continuous Phase

The continuous flow field in the circular bend is simulated using RANS and LES as it is described in Section (6.2.1). Fig. **(6.9)** shows the pressure field, mean velocity and secondary streamlines in the symmetry plane (x=0) as they were predicted by RANS. LES predictions of the same quantities are depicted in Figs. **(6.10)** and **(6.11b)**. Fig. **(6.11a)** shows a LES instantaneous velocity field at a randomly chosen instant. For comparison, instantaneous velocity field and secondary streamlines of the DNS-like work of Breuer *et al.* [3] are presented in Fig. **(6.12)**.

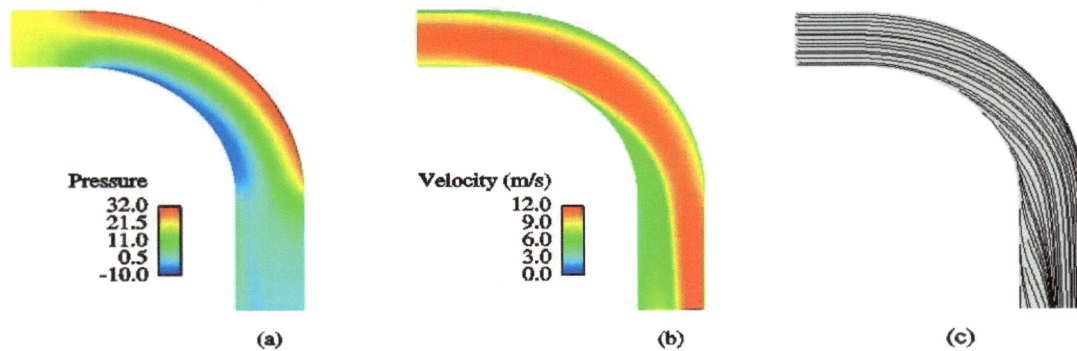

Figure 6.9: RANS predictions of the turbulence flow in the mid-plane of the bend. (a) pressure field of the mean flow, (b) velocity magnitude of the mean flow, (c) Streamlines of the mean flow.

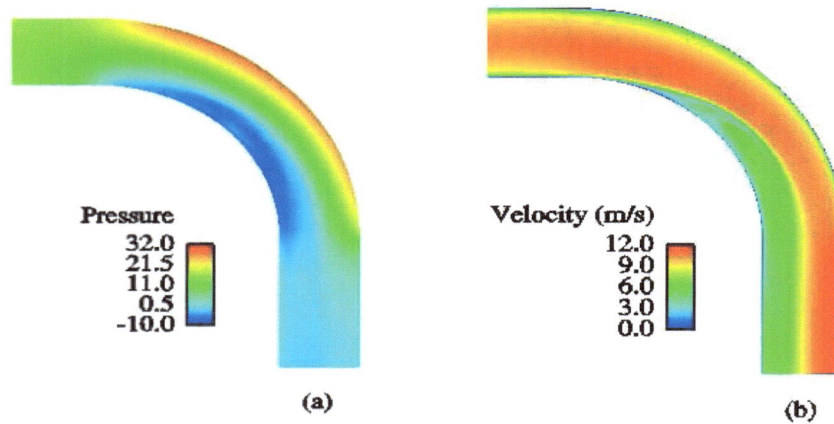

Figure 6.10: LES predictions of the turbulence flow in the mid-plane of the bend. (a) Pressure field of the mean flow, (b) velocity magnitude of the time-averaged flow.

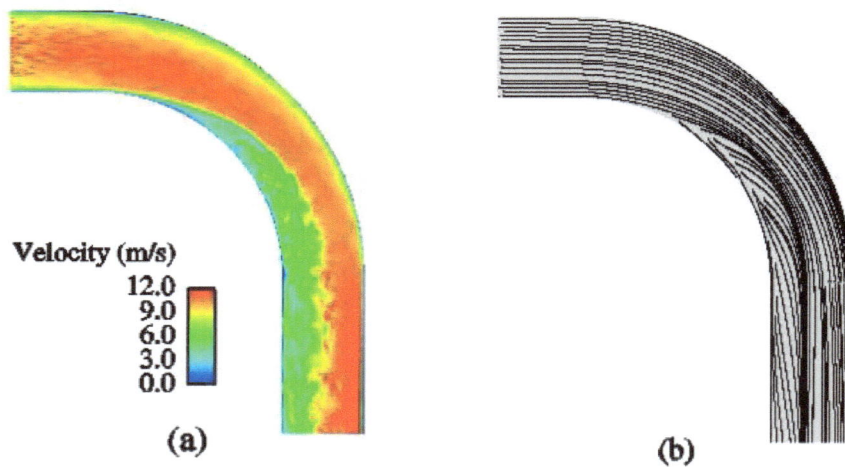

Figure 6.11: LES predictions of the turbulence flow in the mid-plane of the bend. (a) Velocity magnitude of the instantaneous flow, (b) streamlines of the mean flow.

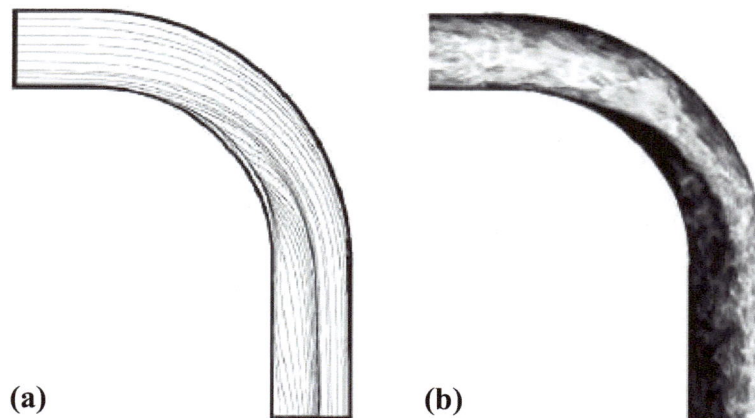

Figure 6.12: Breuer *et al.* [17] predictions of the turbulence flow in the mid-plane of the bend. (a) streamlines of the mean flow, (b) velocity magnitude of the instantaneous flow.

As the fluid flows through the straight pipe and redirected into a bend the pressure which is in the straight section uniform across the flow, must adjust in the bend to counter centrifugal forces. The pressure is greatest at the outer wall farthest from the centre of curvature and least at the inner wall nearest the center of curvature. At the bend inlet the negative pressure gradient on the inner wall is approximately twice the positive gradient on the outside wall. This fact is depicted in Figs. **(6.13)** and **(6.14)** for RANS and LES predictions of pressure field near the wall of the bend entrance respectively. These initial gradients resulting from the change from straight to curved flow disappear at approximately 20° deflection (see Fig. **6.1** for the definition of angle of deflection) so that in the mid section of the bend between 20° and 80° a quasi equilibrium condition is reached with approximately uniform pressure on the inner and outer walls. The transition to straight flow downstream is signaled by the appearance of strong gradients around the 80° deflection. The cross-stream pressure gradient established in the bend as it is depicted in Figs. **(6.9a)** and **(6.10a)** has well known effects on the flow. At the bend inlet, the boundary layer on the outer wall experiences the effect of a positive streamwise pressure gradient which may in a tight bend be sufficiently strong to produce local separation. Conversely the inner wall boundary layer is accelerated. The reverse occurs at the bend at the pipe exit where local pressure gradients of the opposite sign appear as the flow adjusts to uniform pressure conditions downstream.

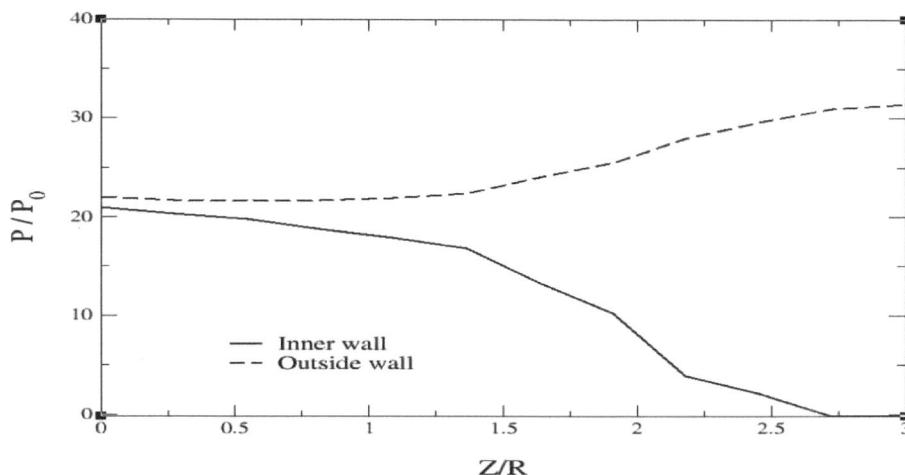

Figure 6.13: RANS predictions of the pressure field near the wall at the bend entrance. P_0 is the reference pressure. Z is the streamwise direction non-dimensionalised by the bend raduis R.

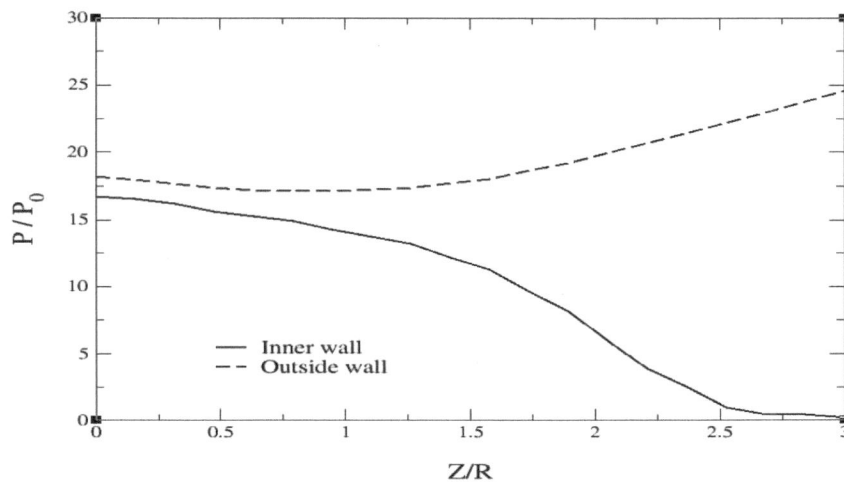

Figure 6.14: LES predictions of the pressure field near the wall at the bend entrance. P_0 is the reference pressure. Z is the streamwise direction non-dimensionalised by the bend radius R.

The impact of the curved geometry on the straight sections usually extends some distance upstream of the bend. For the time-averaged velocity field as portrayed in Figs. **(6.9b)** and **(6.10b)**, it is obvious that the effect of the bend on the flow upstream, *i.e.*, in the straight-pipe, is more pronounced for the RANS results with the flow moving toward the inner side almost at half of the straight pipe length before reaching the bend entrance. This is not the case for the LES predictions though the impact of the curved geometry into the straight section extends some short distance upstream of the bend. The same trend is noticed also on a LES instantaneous velocity field depicted in Fig. **(6.11a)**. It shows also the turbulent structures that appear in the flow. Therefore the resulting flow at the entrance of the bend already differs considerably from a fully developed pipe flow. This should affect the secondary flow patterns within the bend as we shall discuss later.

The streamlines in the plane of symmetry as predicted by RANS and LES calculations are presented in Figs. **(6.9c)** and **(6.11b)** respectively. They provide an indication of the secondary flow in the bend. LES results for the instantaneous velocity field and streamlines compare very well with the predictions of the DNS-like results of Breuer *et al.* [3] as portrayed in Fig. **(6.12)**.

Figs. **(6.15a)** and **(6.15b)** show the turbulent kinetic energy distribution in the mid-plane of the bend as it is predicted by RANS and LES respectively. For both results, the turbulent kinetic energy is amplified on the outer side or "pressure side" of the bend. The velocity magnitude is decreasing as one moves closer to the wall, *i.e.* when the curvature ratio is increasing, and this combined with centrifugal acceleration leads to an unstable situation. On the inner side or "suction side" we have the opposite effect that is a stable situation and turbulence damping. This can be illustrated as follows: if a high speed particle moves toward the wall it encounters lower speed mean flow which is in equilibrium with a lower mean pressure gradient toward the curvature centre. As a result this high speed particle is subject to a centrifugal acceleration that exceeds the pressure gradient and as a result it is pushed back to its equilibrium radial position. This effect is accurately represented by the production terms of the second moment closure model, hence the fairly good agreement between the RANS and LES.

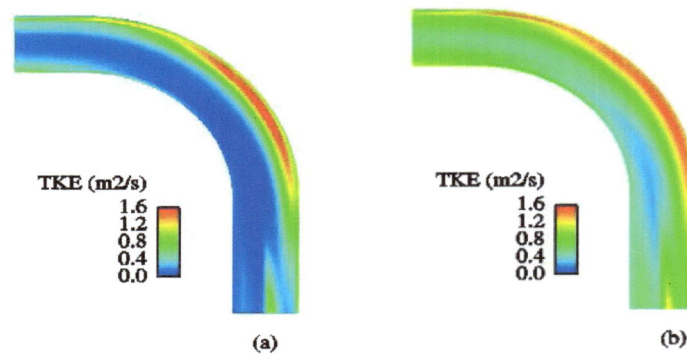

Figure 6.15: Predictions of the turbulent kinetic energy (TKE) in the mid-plane of the bend. (a) RANS results, (b) LES results.

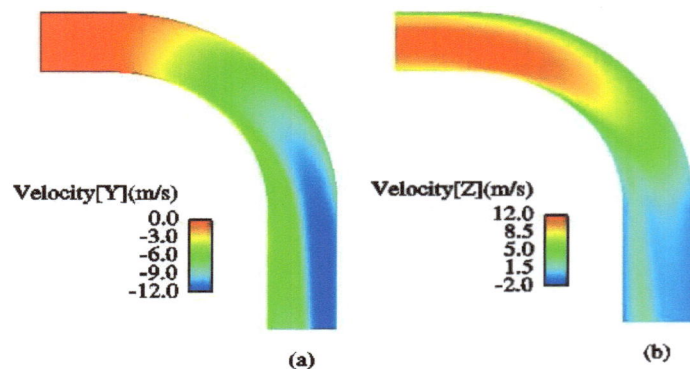

Figure 6.16: RANS predictions of the turbulence flow in the mid-plane of the bend. (a) mean velocity in Y direction, (b) mean velocity in Z direction

Figs. **(6.16)** and **(6.17)** show the mean velocity in Y and Z directions in the plane of symmetry for RANS and LES respectively. For the Y component, LES results show a small zone of recirculation at the inner bend near the entrance that is not predicted by RANS. For Z component, the difference between RANS and LES results are noticed at the bend exit. LES predicts a flowing fluid from the core region toward the inner wall while RANS predicts the contrary, *i.e.*, fluid moving from the walls towards the core region. It is expected that recovery from the bend takes a long distance since it takes a long distance for viscous dissipation to destroy the extra turbulence energy produced by bend curvature [14].

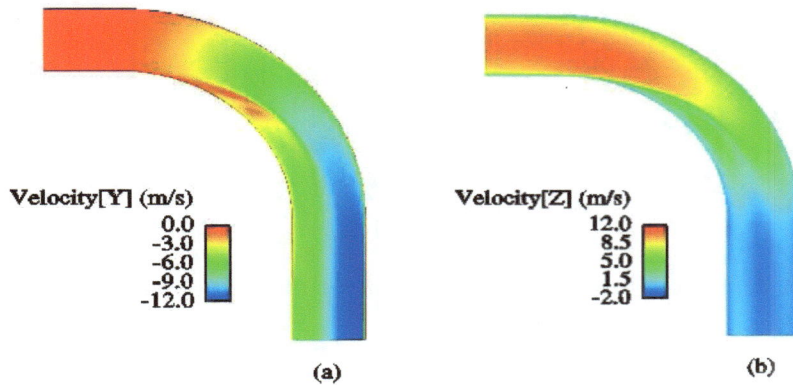

Figure 6.17: LES predictions of the turbulence flow in the mid-plane of the bend. (a) mean velocity in Y direction, (b) mean velocity in Z direction.

As mentioned earlier, the flow in a bend is influenced by centrifugal force due to the curvature. This centrifugal force is, in principle, balanced by a pressure gradient in the plane of curvature. However, near the wall where the velocity is small, this pressure gradient can no longer be balanced and consequently the fluid in the middle of the pipe moves outward and impinges on the outer wall and then turns to move inward along the wall to merge at the inner wall. This flow impingement on the outer wall and separation at the inner wall make flows in curved pipe whether laminar or turbulent very complex. The result is a secondary flow superimposed on the main flow in the plane perpendicular to the mean flow. The magnitude and the shape of this secondary motion depend on the Dean number.

The direct effect of the secondary flow is to displace the region of maximum velocity from the centre towards the outer wall as it is shown in Figs. **(6.18)** and **(6.19)** for RANS and LES respectively. For the bend entrance, the mean axial velocity profile is significantly altered with respect to the fully developed profile in the straight pipe. The location of the maximum velocity is shifted toward the inner bend (I) as it is depicted in Figs. **(6.18a)** and **(6.19a)** for RANS and LES respectively. This is explained by the fact that no centrifugal forces due to the redirection of the flow are present at the bend entrance, but the radial pressure gradient of the curved section is already perceptible as we mentioned earlier. This induces a secondary flow directed to the inner side over the entire cross section.

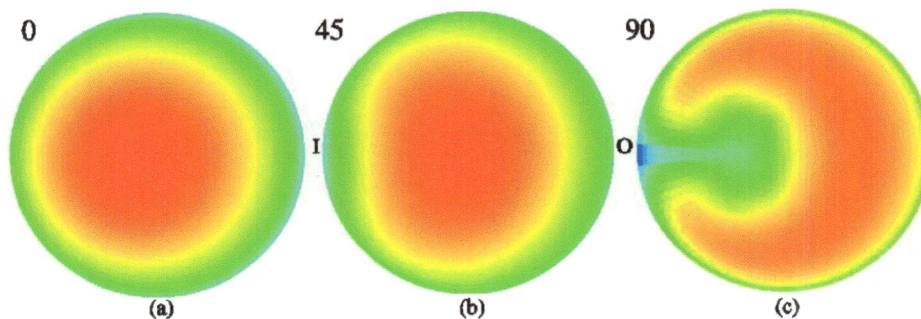

Figure 6.18: RANS predictions of mean axial velocity in different sections of the bend. (a) bend entrance, (b) **45°** deflection, (c) **90°** deflection.

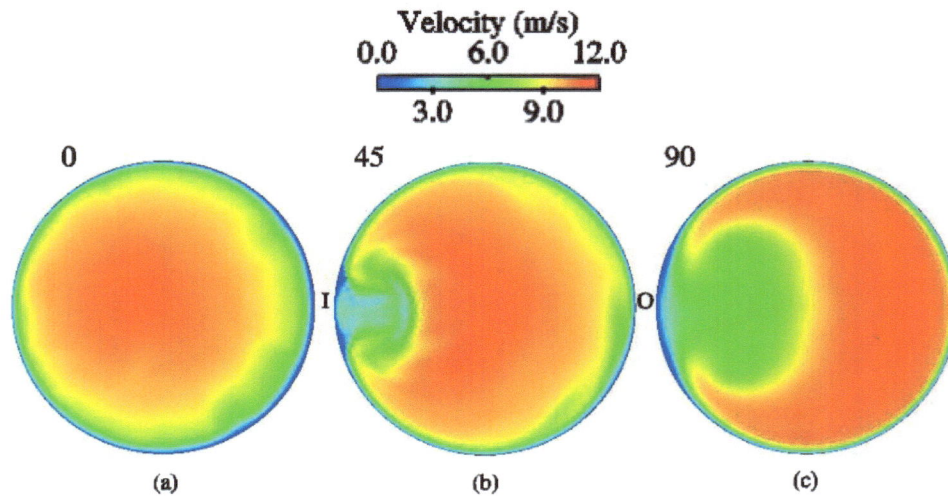

Figure 6.19: {LES predictions of mean axial velocity in different sections of the bend. (a) bend entrance, (b) **45°** deflection, (c) **90°** deflection.

The flow in 90° bends is almost always a developing flow in which the velocity distributions do not attain forms that are independent of the position along the pipe axis. Figs. **(6.18b)** and **(6.18c)** show the cross-section mean axial velocity as it develops downstream the bend entrance at 45° and 90° deflections for RANS calculations. At 45° deflection, the centrifugal forces should become significant but surprisingly the maximum of the mean axial velocity is still similar to the distribution at the entrance of the bend. A thickening of the boundary layer is hardly perceptible at the innermost of the bend. Further downstream at 90° deflection, the maximum of the mean axial velocity has shifted toward the outer radius (O) and low speed boundary layer fluid is now transported from the inner bend to the centre.

The flow development within the bend as predicted by RANS is compared to the LES predictions depicted in Figs. **(6.19b)** and **(6.19c)**. At 45° deflection, the outward movement of the location of the maximum axial velocity is evident in LES results with a perceptible thickening of the secondary boundary layer and zone of flow recirculation around the innermost of the bend. At the bend exit, the time-averaged axial velocity moves further toward the outer radius and the boundary layer is uniformly thick on the inner bend while becoming thinner on the outer radius. These trends are also visible in Figs. **(6.20)** and **(6.21)** depicting the mean axial velocity profile within the bend as predicted by RANS and LES respectively. The secondary motion can be seen more clearly by observing the mean streamlines in cross-sections as it is depicted in Figs. **(6.22)**, **(6.23)** and **(6.24)** for RANS, LES and DNS-like work of Breuer *et al.* [3] respectively.

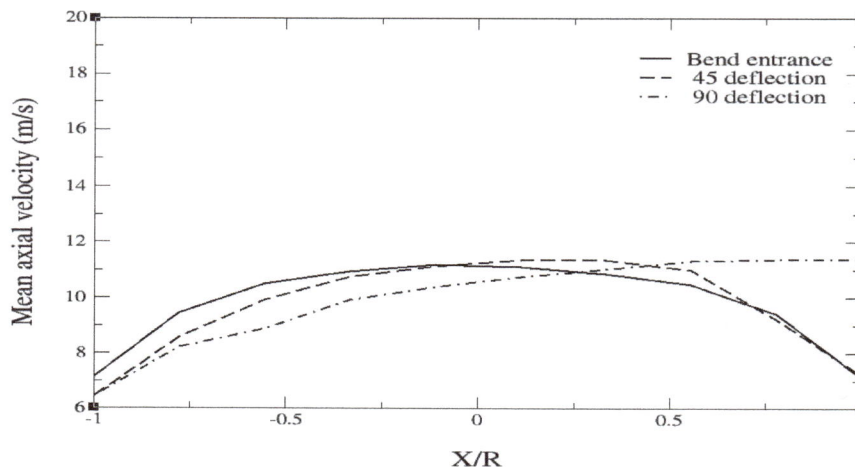

Figure 6.20: RANS predictions of the mean axial velocity profiles for different cross sections within the bend.

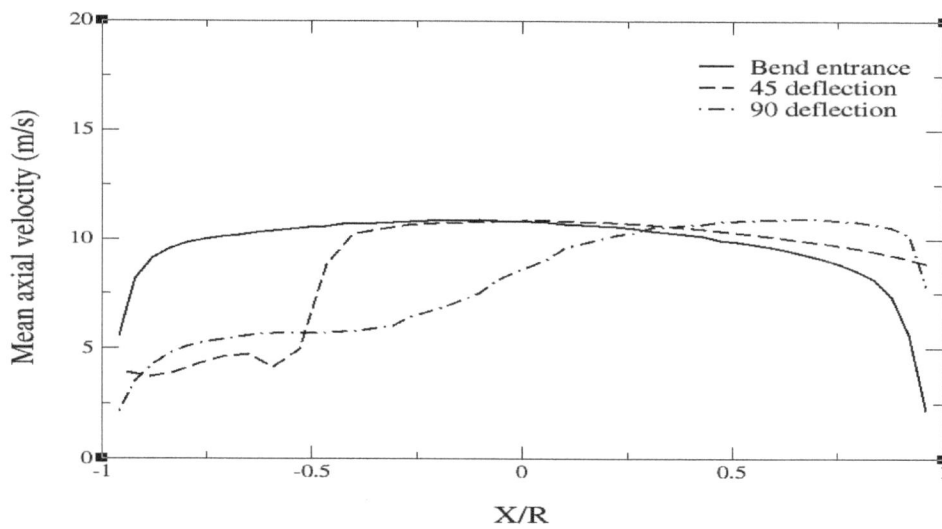

Figure 6.21: LES predictions of the mean axial velocity profiles for different cross sections within the bend.

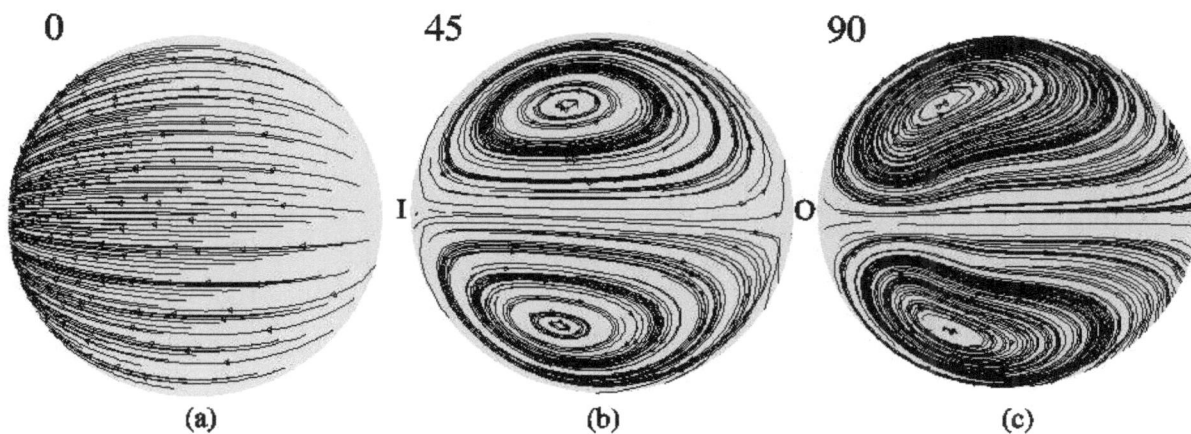

Figure 6.22: RANS predictions of the streamlines of the secondary flow in different sections of the bend. (a) bend entrance, (b) **45°** deflection, (c) **90°** deflection.

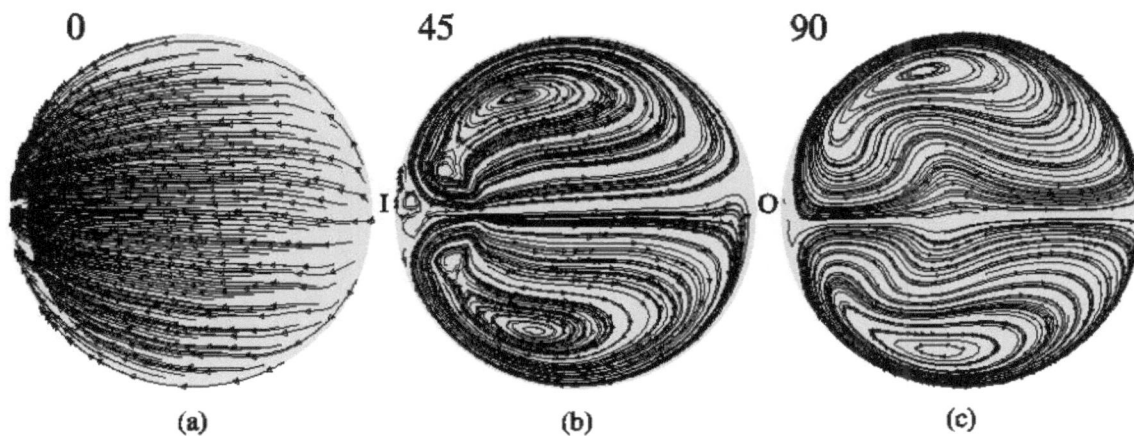

Figure 6.23: LES predictions of the streamlines of the secondary flow in different sections of the bend. (a) bend entrance, (b) **45°** deflection, (c) **90°** deflection.

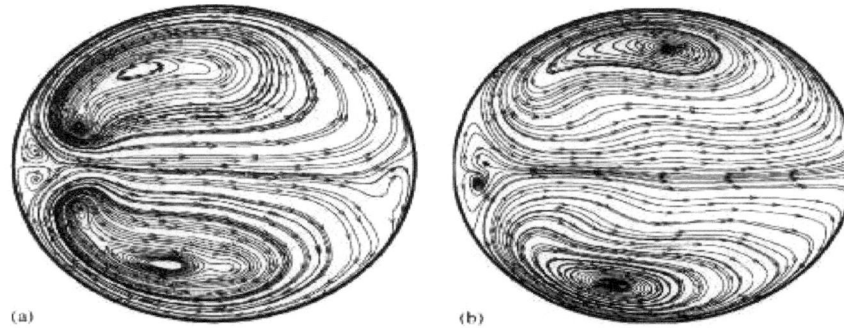

Figure 6.24: Breuer *et al.* [17] predictions of the streamlines of the secondary flow in different sections of the bend. (a) **45°** deflection, (b) **90°** deflection.

At the bend entrance and as mentioned earlier the centrifugal forces are very weak to balance the pressure gradient which results in an inward flow as it is shown in Figs. **(6.22a)** and **(6.23a)**. With increasing deflection, the centrifugal forces increase and the time-averaged flow fields show the well known counter-rotating Dean vortices that circulate in the outward direction in the central part of the pipe. This is the opposite direction compared with the bend entry. This fact is depicted in Figs. **(6.22b)** and **(6.23b)** for the 45° deflection. A comparison between RANS and LES calculations shows that the Dean vortices predicted by LES are more skewed and closer to the wall than the ones predicted by RANS. Moreover, an additional secondary flow structure consisting of a pair of small counter-rotating vortices is predicted by LES and visible at the inner radius in Fig. **(6.23b)**. This pair of weak vortices is not predicted by RANS. At higher deflection, the center of these secondary vortices moves from the inner radius to the outer radius and get closer to the wall as it is shown in Fig. **(6.23c)** for LES calculations. For RANS predictions, Fig. **(6.22c)** show that the pair of the Dean vortices move toward the inner bend instead moving outward without being much skewed. Predictions of the secondary motion for the present LES are compared with the results of Breuer *et al.* [3] depicted in Fig. **(6.24)**. Good agreement is found except in the core region of the cross-section at 90° deflection where a more pronounced distortion of the secondary streamlines is noticed for the present LES.

Comparison of the present LES results with the DNS-like predictions of Breuer *et al.* [3] is quite satisfactory. This gives us confidence in the results presented for which no experimental data are available. Indeed, most of the experimental works investigated turbulent flows in curved bend of a very high Dean number well beyond the Dean number of the test case in hand [15]. For the few works where a comparable Dean number bend flow is studied, the curvature ratio turned out to be less than five. It is well known that in such cases the similitude between flows in curved bend depends on both Reynolds number and curvature ratio and no longer on their combination, *i.e.*, Dean number [16]. Moreover, the work of Pui *et al.* [1] does not provide any results about the continuous phase .

To asses *a-posteriori* the quality of the present LES, the ratio of the subgrid scale kinetic energy k_{SGS} to the total kinetic energy k_T is computed. Fig. **(6.25)** shows an estimation of this ratio in the different directions. This form of presentation gives an overall impression and was chosen since the flow in curved bend does not own a direction of statistical homogeneity. The residual or SGS kinetic energy is estimated using Equations (4.5) and (4.6) in Chapter (4). In X direction and thanks to the non-conforming embedded refinement near the wall, the filtered out kinetic energy is kept under 5% which means that the near-wall turbulence is well resolved. This has a positive repercussion on the prediction of particle deposition that is significantly influenced by the near-wall turbulence. In Y and Z directions, a ratio higher than 20% is noticed in the region $-3 < Y/R < -2$ and $5.5 < Z/R < 6.5$, which is the region around 45° deflection. It is expected that the grid might be not fine enough to capture the dynamics of the separation occurring in that region of the bend. Globally, the ratio of the SGS kinetic energy to the total kinetic energy is well under 20° almost every where in the domain. Consequently the present LES is adequate according to the index of LES quality developed by Celik *et al.* [17].

The failure of RANS to predict correctly the secondary motion patterns can be attributed mainly to the numerical treatment of the near-wall region. It appears that the treatment of the boundary layers using the wall-function procedure has proven inefficient in cases where boundary layers thin in regions and become thick in others and eventually separate. Indeed in the case of turbulent curved bend, the secondary motion is shaped by the behavior of the two boundary layers on the upper and lower halves of the bend. They collide at the innermost point of the cross section, separate there and form a reentrant jet that moves outwards through the core. LES by accurately solving the near-wall turbulence has succeeded in capturing most of the secondary motion patterns as witnessed by Figs. **(6.19)** and **(6.23)**. This has a significant impact on the LES and RANS results concerning particle deposition as we shall discuss below.

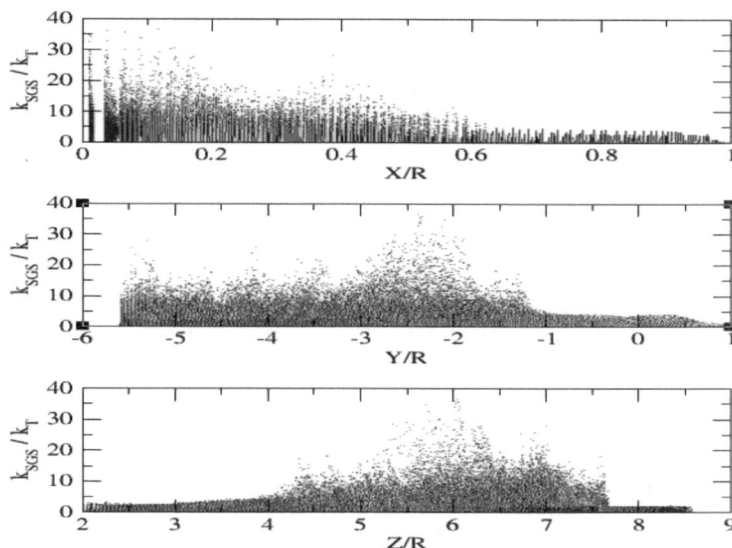

Figure 6.25: Ratio of the subgrid scale kinetic energy k_{SGS} to the total kinetic energy k_t in different directions.

Table 6.3: Physical Characteristics of Inertial Particle Tracked in the Turbulent Bend Flow.

Mean diameter $d_p(\mu m)$	Cunningham factor C_n	Response time $\tau_p(ms)$	Stokes number St
0.62	1.280	0.0011	0.001
4.77	1.042	0.055	0.05
6.80	1.024	0.11	0.1
9.65	1.019	0.22	0.2
11.23	1.015	0.29	0.27
11.64	1.015	0.32	0.29
14.70	1.007	0.5	0.46
18.83	1.003	0.82	0.75
19.95	1	0.92	0.84
21.77	1	1.09	1
24.05	1	1.33	1.22
25.30	1	1.47	1.35
26.66	1	1.64	1.5

6.3.2 Dispersed Phase

For the dispersed phase, the deposition efficiency is computed for different particle Stokes numbers using Equation (6.1). The Stokes number of the liquid particles considered ranges from 0.001 to 1.5 as it shown in Table **(6.3)**. This range of particle Stokes number covers a wide inertial particle behavior towards the turbulence SGS present in the flow. For particles with small Stokes number ($St < 0.2$) and as far as LES calculations are concerned, it is expected that the subgrid scale turbulence should have a significant impact on particle transport and thus deposition. RANS is expected to fail to predict the deposition efficiency of particles with Stokes numbers falling within this range. This was the case when the dispersion of small-Stokes-number was investigated in Chapter 5. For Particle with intermediate Stokes number ($0.2 < St < 1$), they should be rather transported by the resolved turbulence with a possibility of SGS turbulence to influence particles with Stokes number at the lower end of this range. For the third category that contains particles with Stokes numbers equal or larger than 1, they will not sense the turbulence present in the flow and their inertial deposition will be affected by the mean flow. For these particles, RANS and LES using only the resolved field are therefore expected to predict well their deposition efficiencies.

Figs. (6.26) and (6.27) depict the predicted RANS and LES results respectively concerning the deposition efficiency. They are compared to the measurements of Pui *et al.* [1] and the DNS-like work of Breuer *et al.* [3]. Furthermore, a curve fitted through the experimental data by Pui *et al.* [1] is added which describes the deposition efficiency as a function of the Stokes number,

$$\eta_p = 1 - 10^{-0.963St}. \tag{6.17}$$

It is worthy to mention that relation (6.17) is a fit for the experimental observations of Pui *et al.* [1] for particles with Stokes number between 0.27 and 1.35. Therefore the curve is displayed in dashed line for $St < 0.27$.

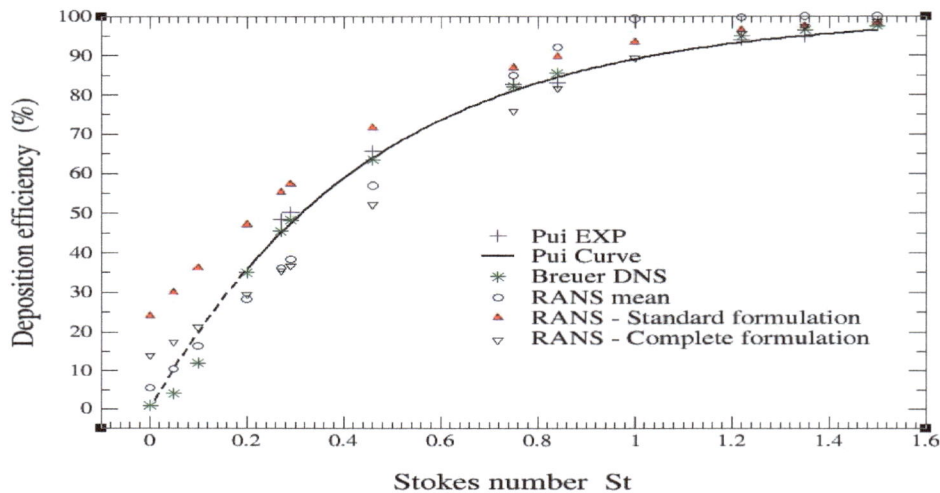

Figure 6.26: RANS predictions of deposition efficiency versus particle Stokes number.

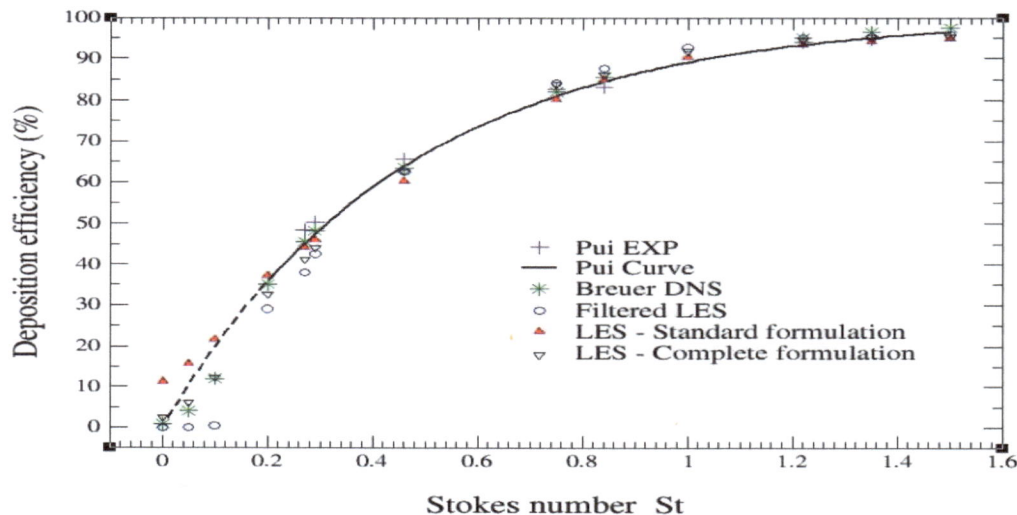

Figure 6.27: LES predictions of deposition efficiency versus particle Stokes number.

RANS results for particle deposition are shown in Fig. **(6.26)**. First, particles are tracked using only the mean flow generated by RANS. Results show an over-prediction of the deposition efficiency for the small-Stokes-number particles and for particle with Stokes number higher than 0.75. Particles deposit on the walls with a 100% of efficiency as early as Stokes number equal to 1. For particles with intermediate Stokes number, the deposition efficiency is underestimated. Including the turbulence fluctuations through both standard and complete formulations of the stochastic model does not yield any amelioration of the results in particular for particles with Stokes number smaller than 1. The standard formulation appears to generate more subgrid turbulence than it should and this is witnessed by an overestimated deposition efficiency over the whole range of Stokes numbers considered. The complete formulation provides better results compared to the standard formulation in particular for particles with Stokes number higher than 0.75.

As it was mentioned in Chapter 5, in the framework of RANS the model is unable to reproduce with certain accuracy the whole spectrum of turbulence that is highly anisotropic and characterized by a wide range of length and time scales.

Figs. **(6.27)** depicts the LES results of deposition efficiency. The use of the filtered velocity field only to track the inertial particles with Stokes numbers higher than 0.5 shows good agreement with the experimental results of Pui *et al.* [1] and the DNS-like work of Breuer *et al.* [3]. For Stokes number $St < 0.5$, the deposition is underestimated and the differences between the LES predictions and the reference results become significant as the Stokes number becomes smaller. This suggests that for these particles, the large-scale turbulence fluctuations contained in the filtered LES velocity field are not solely responsible of the dispersion and deposition of particles with Stokes numbers less than 0.5. Thus, the discarded SGS turbulence should be taken into account in order to predict the correct level of turbulence these particles should see and thus the correct deposition efficiency.

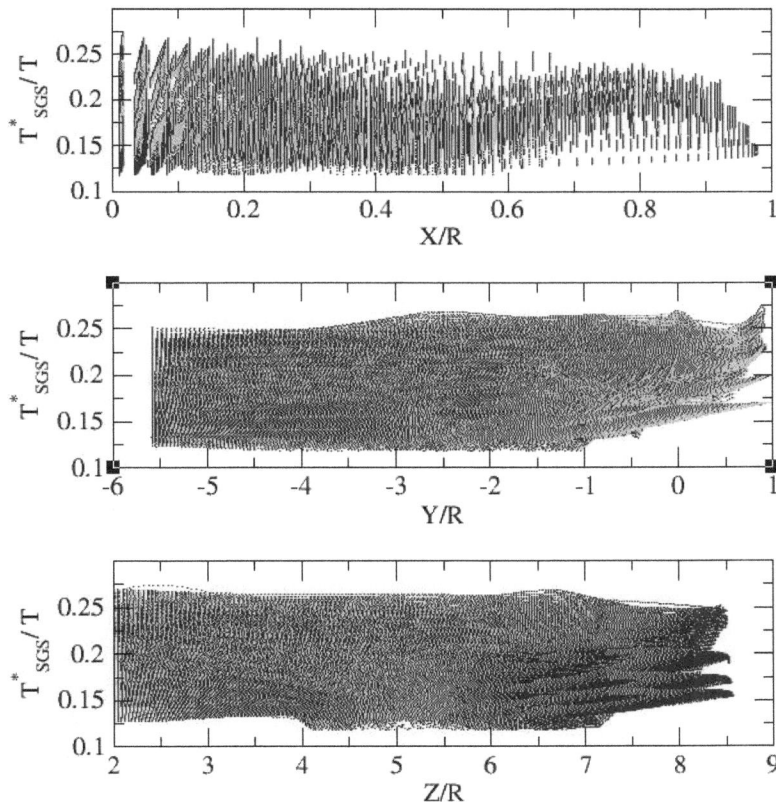

Figure 6.28: Ratio of the subgrid time scale T^*_{SGS} to the integral time scale T in different direction. $T = R/U_0$

The fluid velocity seen by the particles taking into account SGS fluctuations is constructed as it is described in An estimation of the time scale of the SGS fluctuations that is washed out by the filtering operation in the present LES shows that the SGS motion should have an important role for the dispersion and deposition of part of the particles considered in theses simulations. Fig. **(6.28)** presents an estimation of the SGS time scales in the computation domain in different directions. The SGS time scale T^*_{SGS} is non-dimensionalised by the integral time scale $T = R/U_o$ on which the particle Stokes number in Figs. **(6.26)** and **(6.27)** is based. This estimation is made according to Equations (4.9) in Chapter **4**. For the case in hand the SGS time scale ranges between 0.12T and 0.25T. Consequently particles with response times smaller than 0.25T will sense a portion or the whole turbulence linked to the SGS motion depending on their response times. This agrees well with the results of Fig. **(6.27)** where a deviation between the deposition efficiency as predicted by LES using only the filtered field and the reference results becomes noticeable for particles with a response time equals to 0.29T and become more significant as the Stokes number goes down.

Chapter **4** using both standard and complete formulations. Results of the deposition efficiency using LES with the complete formulation of the stochastic model show a very good agreement with the DNS-like work of Breuer *et al.* [3]. They slightly overestimate η_p for the small-Stokes-number particles ($St < 0.1$) and slightly underestimate it for particles with Stokes number between ($0.2 < St < 0.3$). For the standard formulation, it appears that the model generates too much turbulent fluctuations that results in a significant overestimation of the deposition efficiency for the small-Stokes-number particles while for particles with Stokes number between ($0.2 < St < 0.3$) a very good agreement is found. For the particles with Stokes number higher than 0.5, the LES predictions of the deposition efficiency should be similar whether the only filtered velocity is used or when the stochastic model is used because these particles simply do not sense the SGS turbulent fluctuations. Fig. **(6.27)** shows that this is the case, and very good agreement with the reference results is found.

According to the predictions of the present LES and the DNS-like work of Breuer *et al.* [3], relation (6.17) does not hold for the lower Stokes numbers ($St < 0.2$). Indeed, the numerical results of the deposition efficiency versus Stokes number show an inflection point and approach zero for finite Stokes number unlike the exponential relation (6.17) that approaches zero only at St = 0.

The effect of Stokes number on particle deposition behavior is also investigated. Figs. **(6.29)** and **(6.30)** show the LES predictions of deposition patterns on the outer and inner bend respectively for nine different Stokes numbers. The small particles with low inertia are able to follow the continuous flow closely. They are subject to all the turbulent fluctuations that are present in the flow with different time scales. They mostly exit the bend without depositing as it is shown on Figs. **(6.29a)**, **(6.29b)** and **(6.29c)**. A small fraction of them deposits on both sides of the wall or on the inner bend owing to the secondary motion as depicted in Figs. **(6.30a)**, **(6.30b)** and **(6.30c)**.

Particles with intermediate Stokes number are more subject to inertia forces owing to their increased sizes. They respond to less and less turbulent fluctuations because of their growing Stokes numbers. Due to centrifugal forces, particles that are in the core region of the bend are entrained by the reentrant jet and thus deposited on the outer bend while particles that are trapped in the counter-rotating Dean vortices exit the bend without depositing. For particles with Stokes number *St*=0.27, a deposition on the outer bend is occurring only on the outermost radius as it is shown in Fig. **(6.29d)**. For particles with Stokes numbers *St*=0.46 and *St*=0.75, the outer bend is completely covered with depositing particles as it is depicted in Figs. **(6.29e)** and **(6.29f)**. Owing to their increased inertia the last two classes of particles cease to follow the flow in the Dean vortices and are driven toward the outer bend by the centrifugal forces. For this range of Stokes number, part of particles that exit the bend deposit on both sides of the straight pipe as it is shown in Figs. **(6.30d)**, **(6.30e)** and **(6.30f)**.

For particles with large Stokes numbers ($St > 1$) they are no longer able to fellow the secondary flow in the bend. So they almost all deposit at the outer bend because of the centrifugal force and only an insignificant number of these particles deposit on the inner bend or exit the bend. This fact is depicted in Figs. **(6.30g)**, **(6.30h)** and **(6.30i)**.

It is noticeable in Figs. **(6.30a)** to **(6.30i)** that there is a long narrow band around the innermost radius where no deposition takes place. This fact is mentioned in Breuer *et al.* work [3] but not reported in the experimental study of Pui *et al.* [1]. The reason for this phenomenon is the additional secondary flow structure consisting of a pair of counter-rotating small vortices at the inner bend radius that prevent particles from depositing along a stripe at a region around the innermost bend.

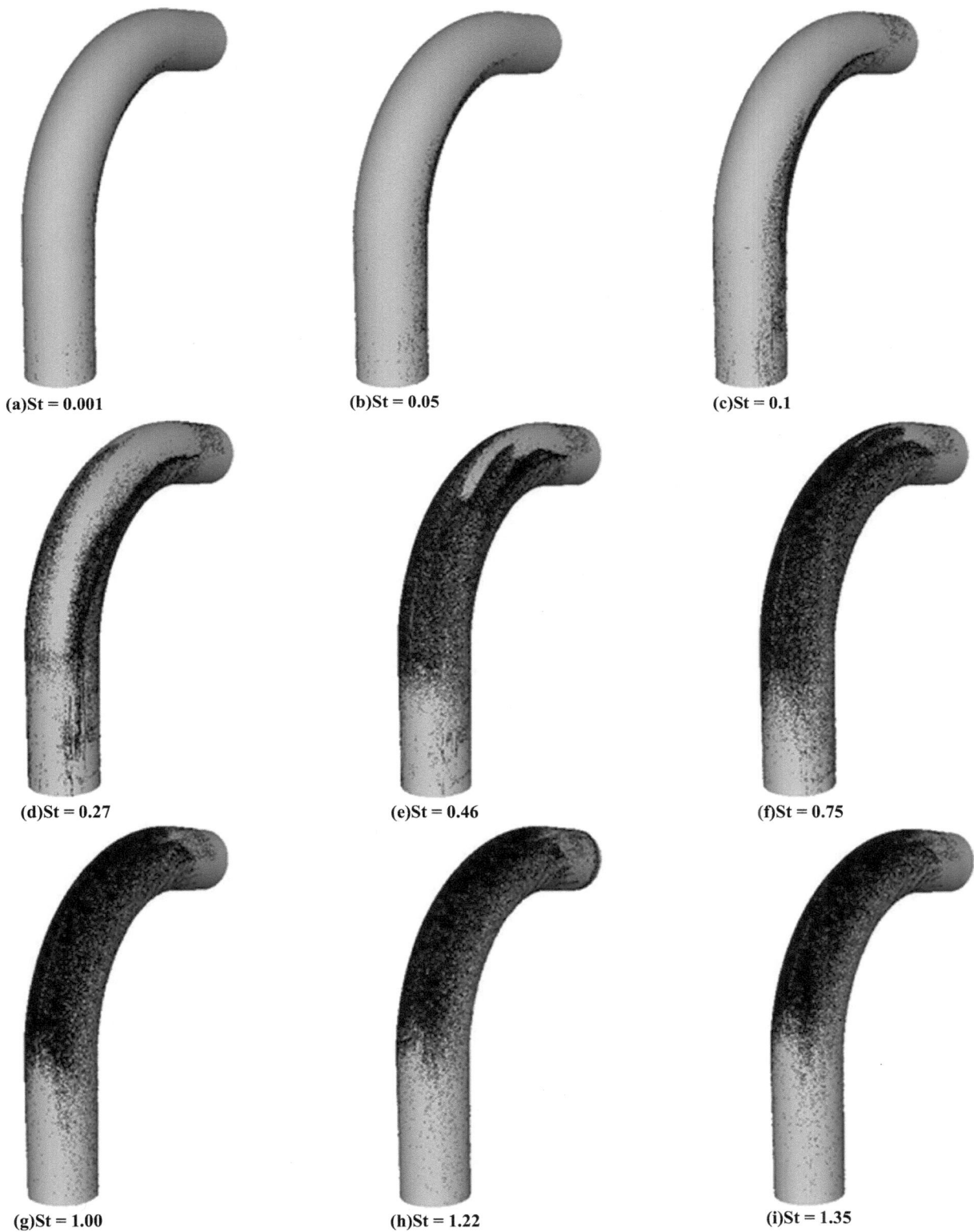

Figure 6.29: LES predictions of particle deposition patterns on the outer bend at nine different Stokes numbers: (a) St = 0.001, (b) St = 0.05, (c) St = 0.1, (d) St = 0.27, (e) St = 0.46, (f) St = 0.75, (g) St = 1.00, (h) St = 1.22, (i) St = 1.35.

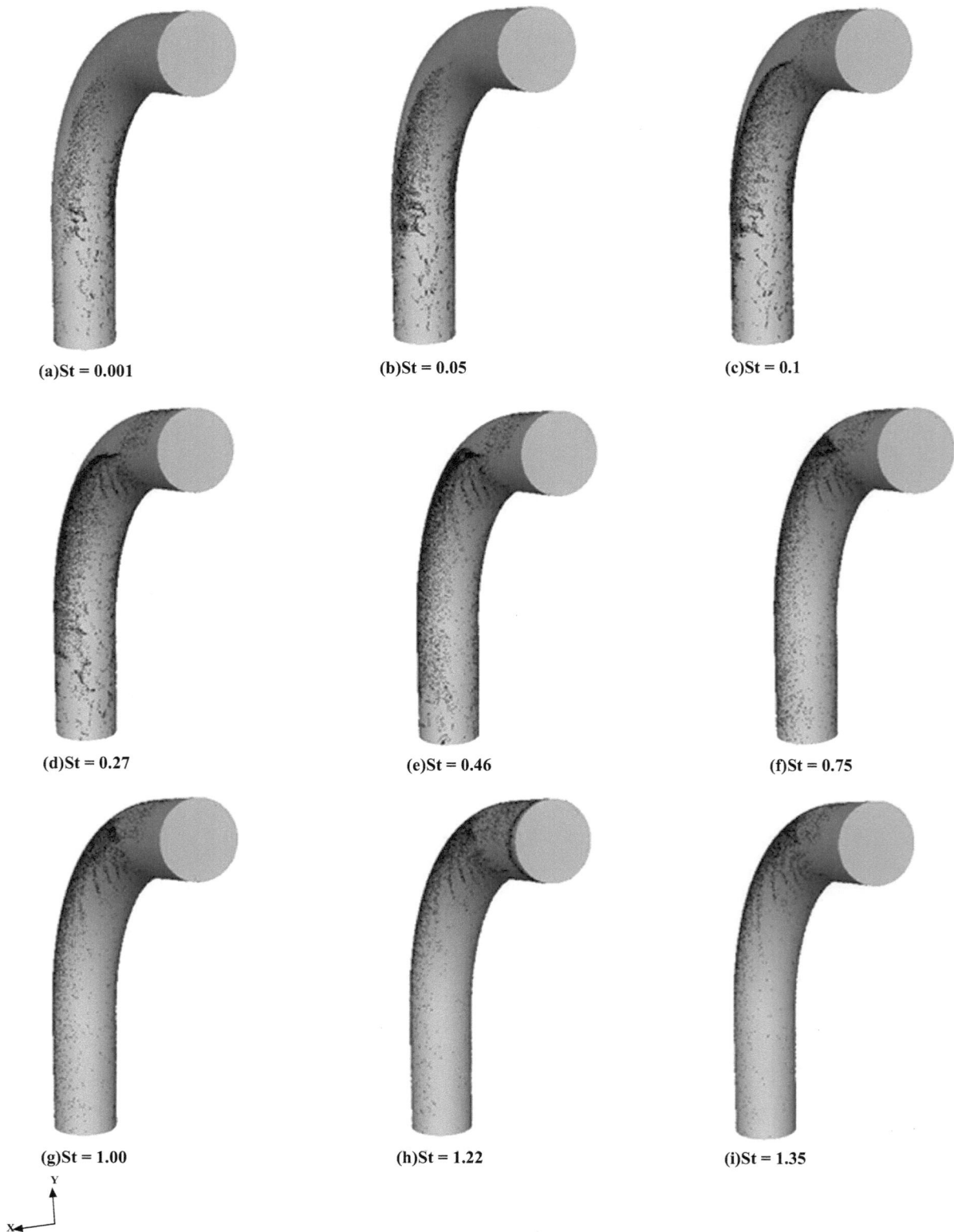

Figure 6.30: LES predictions of particle deposition patterns on the outer bend at nine different Stokes numbers: (a) St = 0.001, (b) St = 0.05, (c) St = 0.1, (d) St = 0.27, (e) St = 0.46, (f) St = 0.75, (g) St = 1.00, (h) St = 1.22, (i) St = 1.35.

To investigate the effect of particle injection position on the deposition efficiency, the section of injection is divided into 5 regions or bins ($N_b = 5$) of equal surface area as it is shown in Fig. **(6.31)**. Particles are injected randomly at the injection section and the number of particles originated from the same region and deposited on the bend wall is computed and displayed in Fig. **(6.32)**. For particle with Stokes number $St = 0.001$, no particle originating from zone 1 and 2 that represent the core region of the bend, deposit on the bend walls. Around 60% of particles that came into contact with the bend walls are those that were injected near the wall from zone 5. The contribution of this zone to the number of depositing particles decreases with increasing Stokes number while more and more particles injected from zone 2 and 3 deposited. It is interesting to notice that particles with Stokes number up to $St = 0.1$ do not deposit if they are injected from zone 1.

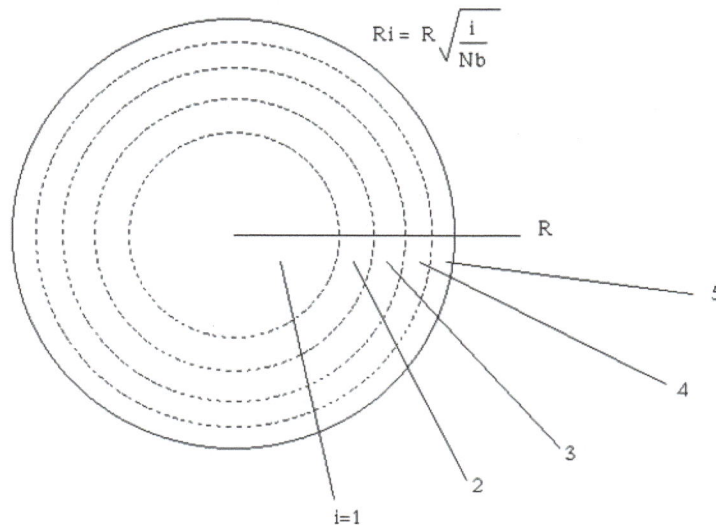

$$R_i = R \sqrt{\frac{i}{N_b}}$$

Figure 6.31: Definition of equal area intervals on the section of injection.

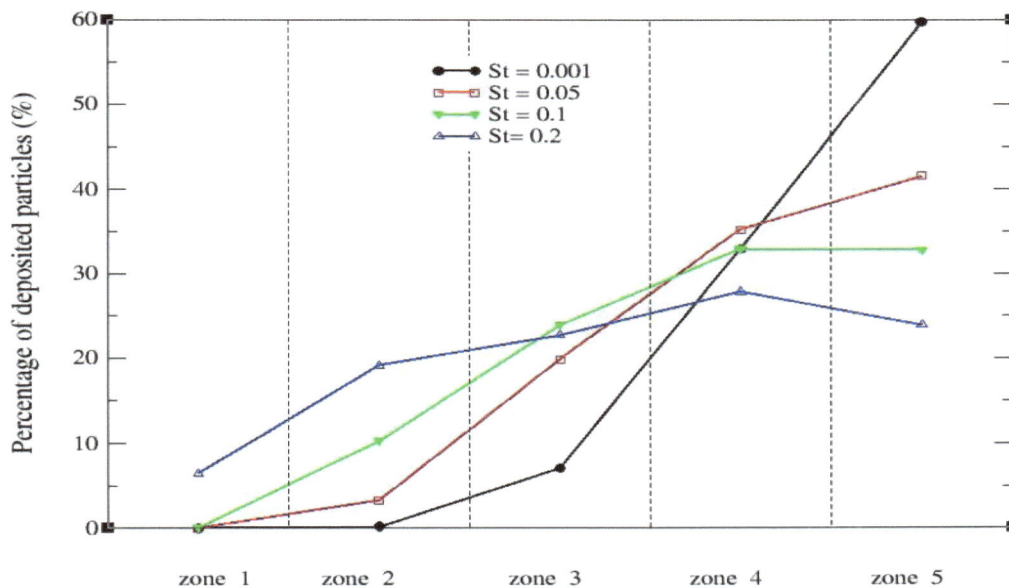

Figure 6.32: Percentage of deposited particle per zone of injection defined in Figure (6.31)

This remark could have an important practical use concerning the design of the Metered Dose Aerosol Inhalers (MDI) that are used to deliver medication through the mouth to the lungs for people with respiratory system diseases. It is well known that the smaller are the aerosols generated by these MDIs, the lower is their deposition rate in the mouth and throat and by consequence the less are the drug's side effects. At the same time the smaller are the aerosols the more expensive the device is. The flow parameters that are studied herein and the geometry in which it takes place are similar to the flow in the mouth-throat geometry. Consequently, the main findings of this study can be used to design a cheap and efficient MDI. Indeed the results show that if particles are injected on the whole cross section of the MDI's section of injection, particles with diameter as small as $d_p = 0.62 \mu m$ ($St = 0.001$) should be generated in order to have an insignificant deposition in the mouth-throat region. However, if the injection is confined to the core region of the MDI's section of injection (as can be obtained by a co-flow device), insignificant deposition occurs for particle diameters as large as $d_p = 6.8 \, \mu m$ (St=0.1).

REFERENCES

[1]   Pui DYH, Romay-Novas F, and Liu BYH. Experimental study of particle deposition in bend of circular cross section. Aerosol Sci Technol 1987; 7; 301-305.

[2]   Cheng YS, Wang CS. Motion of particles in bends of circular pipes. Atmos Environ 1981; 15; 301-306.

[3]   Breuer M, Baytekin HT, Matida EA. Prediction of aerosol deposition in 90 bends using LES and an efficient Lagrangian tracking method. J Aerosol Sci 2006; 37; 407-1428

[4]   Stapleton KW, Guentsch E, Hoskinson MK, Finlay WH. On the suitability of $k - \epsilon$ turbulence modeling for aerosol deposition in the mouth and throat: a comparison with experiment. J Aerosol Sci 2000; 31(6); 739-749.

[5]   Eggels JGM, Boersma BJ, Nieuwstadt FTM. Direct and large-eddy simulations of turbulent flow in an axially rotating pipe. In Proceeding of the 9th Symposium on Turbuelnt Shear Flows 1993; pp. 301

[6]   Durbin PA, Pettersson Reif BA. Statistical theory and modeling for turbulent flows. Chichester: John Wiley & Sons 2003.

[7]   Wilcox DC. Turbulence modeling for CFD. DCW Industries Inc., La Canada, CA 1993.

[8]   Launder BE, Reece GJ, Rodi W. Progress in the development of a Reynolds-stress turbulent closure. J Fluid Mechan 1975; 537-566 pp.

[9]   Speziale CG, Sarkar S, Gatski TB. Modeling the pressure-strain correlation of turbulence: an invariant dynamical system approach. J Fluid Mechan 1991; 245-272 pp.

[10]   Grotjans H, Mentor F. Wall functions for general application CFD codes. In Papailou et al., editor, ECCOMAS 1998; 98; 1112-1117.

[11]   Oesterle B, Zaichik L. On Lagrangian time scales and particle dispersion modeling in equilibrium turbulent shear flows. Phys Fluids 2004; 16(9); 3374-3384.

[12]   Kleinstreuer C. Two-phase flow: theory and applications. Taylor and Francis Group Inc 2003.

[13]   Wang LP, Stock DE. Dispersion of heavy particles by turbulent motion. J Atmos Sci 1993; 50(13); 1897-1913.

[14]   Enayet MM, Gibson MM, Taylor AMKP, Yianneskis M. Laser-Doppler measurements of laminar and turbulent flow in a pipe bend. Int J Heat Fluid Flow 1982; pp. 213-219.

[15]   Boersma BJ, Nieuwstadt FTM. Large-eddy simulation of turbulent flow in a curved pipe. J Fluids Engine 1996; 118; 248-254.

[16]   Berger SA, Talbot L, Yao LS. Flow in curved pipes. Ann Rev Fluid Mech 1983; 15; 461-512.

[17]   Celik IB, Cehreli ZN, Yavuz I. Index of quality for large-eddy simulations. Proceedings of ASME FEDSM2003-45448. 4th ASME JSME joint Fluids Engineering Conference. Honolulu, Hawaii 2003.

CHAPTER 7

Expiratory Droplet Dispersion in a Mechanically Ventilated Enclosure

Abstract: After the epidemic outburst of avian influenza and severe acute respiratory syndrome (SARS) in East and Southeast Asia, there has been a burgeoning research interest in investigating the control and transport's mechanisms of airborne bacteria and viruses indoors and in confined environments such as aircraft cabin. Dispersion of microorganism-laden aerosols exhaled from infected patients was admitted as a potential airborne transmission pathway. Consequently, competent understanding of aerosol dispersion and deposition is necessary to improve exposure assessment tools and models and endorse efficient ventilation strategies that can considerably reduce indoor particle concentrations and improve the indoor air quality.

This investigation attempts to provide a realistic simulation of the time-dependant flow field in a chamber inhabited with two heated manikins using the well resolved LES approach. The potential of both the Eulerian and the Lagrangian approaches coupled to LES of non-isothermal airflow to study dispersion characteristics of expiratory aerosols has been examined.

Experimental and numerical findings on the flow and temperature fields were recorded and show all good agreements. By taking into account the uncertainly of the particle counter used in the experimental work, globally the agreement between the experimental results and the computational modeling predictions on aerosols rates of decay is quite acceptable.

Keywords: Indoor air quality; Large eddy simulation; particle-laden flow; turbulence; Lagrangian description; Eulerian description; aerosols; thermal effects; deposition; ventilation schemes.

7.1 INTRODUCTION

Indoor Air Quality (IAQ) matters have been receiving much research attention from various disciplines over the last couple of decades [1]. This has been mainly triggered by the necessity of protecting indoor environments against the intentional release of chemical and/or biological agents on one hand, and by the need of promoting more comfortable and healthy indoor environment on the other hand.

After the epidemic outburst of avian influenza and severe acute respiratory syndrome (SARS) in East and Southeast Asia, there has been a burgeoning research interest in investigation the control and transport's mechanisms of airborne bacteria and viruses indoors and in confined environments such as aircraft cabin. Dispersion of microorganism-laden aerosols exhaled from infected patients was admitted as a potential airborne transmission pathway. Consequently, competent understanding of aerosol dispersion and deposition is necessary to improve exposure assessment tools and models and endorse efficient ventilation strategies that can considerably reduce indoor particle concentrations and improve the indoor air quality.

Depending on the original and final size, droplet nucleus can remain suspended in air for several hours and thus spread widely throughout indoors. This also depends on the ventilation scheme used. Indeed, the ventilation system determines the airflow pattern in the room which in turn decides the droplet nucleus fate. Displacement ventilation has been acknowledged to ensure better indoor air quality than mixing ventilation. Also, in the presence of heat sources, displacement ventilation was found to be more contributive to pollutants removal without much mixing to the whole indoor environments and this compared to other types of ventilation.

It is well known that in order to design an efficacious ventilation system, it is critical to have a reliable tool that is capable of predicting airflow pattern and particle transport indoors. This can be achieved through the use of Computational Fluid Dynamics (CFD) which has the capacity to provide microscopic information on the indoor air environment like the air velocity, pressure, temperature, and pollutant's concentration distribution which are useful to obtain pertinent macroscopic parameters for engineering goals.

Among these CFD tools, Reynolds averaged Navier–Stokes (RANS) simulation has been widely deployed to simulate airflow indoors and predict the averaged velocity or temperature. Large Eddy Simulation (LES) solves directly for the transient behaviour of the large-scale turbulent motion which tends to have the greatest influence on the turbulent transport with having the advantage of less constants to adjust [2]. LES approach to turbulent flows has been recognised as a powerful tool that is able to satisfy a continuing desire for higher fidelity of predictive capabilities. Its use has been stimulated by the extraordinary capabilities' progress of computational means. Since LES solves time-dependant turbulent flows, it can supply detailed description of the turbulence phenomenon, such as three-dimensional instantaneous velocity field. Thus, LES accurately accounts for the history and transport effects of turbulence on aerosol dispersion and deposition. The very informative picture of the turbulent transport provided by LES has usually been associated with large computational expenses that have prevented its use in the past. However, many recent works have shown that LES computations carried out on current computer workstations have been successfully applied to several indoor airflow problems with reasonable computation costs. Zhang *et al.* [3] tested various turbulence models, including LES, to predict airflow and turbulence in enclosed environments. Zhang and Chen [2] proposed a filtered dynamic subgrid scale model to be used in association with LES of indoor airflow. All these investigations have shown the very promising potential of LES in predicting turbulent airflow in confined environments but none of them have considered thermal effects generated by humans.

For LES of two-phase turbulent flows, the numerical simulation of the particulate trajectories and dispersion pattern in airflows are investigated according to two streams namely the Lagrangian and the Eulerian methods. The Eulerian methodologies treat the particle phase as another continuum. Transport equation for the mass concentration is developed from the mass (species) conservation conditions and solved to detail the particle concentration field. When the particle phase is considered as scalar species, the gravitational settling and the deposition rate should be properly taken into account to reflect the aerosol's inertial character [1]. For the Lagrangian approach, the motion of the discrete particles is described by the force balance including those induced by the interactions with the carrier phase. In indoor particle studies, each method gains its own reputation depending on the research principal aims. The Eulerian method is extensively used to predict particle concentration distributions in rooms. Generally these simulations agree reasonably well with experimental data although noticeable discrepancies exist in some studies [1]. If the particle motion and dispersion history is of interest, the Lagrangian method is more applicable. Indeed, the Lagrangian approach appears to ensure a more natural account of the actual physical phenomena compared to the Eulerian approach. Investigations using two-phase LES to study particle transport in enclosed environments are rare in literature.

Almost all of the previous studies did not include thermal effects. For the common ceiling schemes, the influence of high momentum airflow overwhelms the buoyancy effect and the pollutant transport is mainly dictated by the bulk airflow. Nevertheless the working principle of displacement ventilations relies entirely on the existence of internal heat sources, hence the importance of studying aerosol transport by displacement ventilation with internal heat sources. In the present work, measurements and numerical simulations of contaminant particle concentration evolution in a mechanically ventilated room with heated sources have been carried out.

This investigation attempts to provide a realistic simulation of the time-dependant flow field in a chamber inhabited with two heated manikins using the well resolved LES approach. The potential of both the Eulerian and the Lagrangian approaches coupled to LES of non-isothermal airflow to study dispersion characteristics of expiratory aerosols has been examined. It is important to mention that the experimental and numerical findings presented in this chapter should not be treated as portable and hence draw practical conclusions about the effectiveness of the displacement ventilation scheme for the contaminant removal in full-size model. This is because the scaling rules were not respected, in particular for the thermal aspects due to the physical restrictions imposed by the type of chamber used for the experiment.

7.2 CASE MODEL DESCRIPTION

Both physical and numerical experiments on airflow field and particle concentration were carried out for a downscaled chamber with two identical model occupants. The chamber is mechanically ventilated using the displacement ventilation scheme. Table **(7.1)** shows the details of the room, occupants' geometries, and the boundary conditions. The centre point of each object is indicated the Table **(7.1)**. The geometry of the human occupant used in this study is the one often proposed in literature [2]. An opening (0.004 m × 0.004 m), located at the centreline of the head and measured at 0.33 m above floor was added to simulate the mouth of the occupant. Table **(7.2)** shows the details of the geometry of the model occupant used for this study. Two planes are defined in the geometry; a plane (X–Y) crossing through the human model at the inlet and a mid-plane (Y–Z) at X = 0.25 m. The occupant temperature is set to 317 K while the wall temperature is set to 297 K. One of the occupants is emitting droplets (source) and faces directly the second occupant (receptor). The source emits water spherical droplets with initial velocity of 10 m/s for a time period of 0.1 s; similar duration period has been used in the literature [4].

Table 7.1: Geometry and Boundary Conditions

	Location (m)			Location (m)			T (K)	V (m/s)
	X	Y	Z	ΔX	ΔY	ΔZ		
Room	-	-	-	0.5	0.5	1.0	-	-
Inlet	0.25	0.075	0	0.1	0.1	-	297	0.2
Outlet	0.25	0.5	0.75	0.1	-	0.1	-	-
Manikin (1)	0.25	-	0.4	-	-	-	317	-
Manikin (2)	0.25	-	0.6	-	-	-	-	-

Table 7.2: Geometry of the Model Occupant

Part	Dimension (m)
Head	$D = 0.03, L_Z = 0.04$
Torso	$L_X = 0.10, L_Y = 0.14, L_Z = 0.03$
Leg	$D = 0.03, L_Z = 0.17$
Mouth	$L_X = 0.004, L_Y = 0.004$

In this work, some well-justified assumptions were made. First, the evaporation period of the emitted droplets was not taken into account and only the droplet nuclei are directly modelled. The shrinkage time from the original droplet to droplet nuclei is very short and it is estimated in the order of 0.5 s [2]. It is at least one order of magnitude shorter than the residence time of the droplet nuclei present in the room. Also, the droplets are assumed trapped once they touch any surfaces and will not rebound or break up. These assumptions are genuine for the present low air velocity environment. Coagulation effects have been examined by applying a simple estimation [1] and the results revealed that they can be neglected.

7.3 GOVERNING EQUATIONS OF DROPLET-GAS TURBULENT FLOW

7.3.1 Numerical Description of Airflow

The filtered spatial and temporal evolution of non-isothermal incompressible Newtonian fluid flow is governed by the following equations:

$$\frac{\partial \bar{u}_i}{\partial x_i} = 0 \tag{7.1}$$

$$\frac{\partial \bar{u}_i}{\partial t} + \frac{\partial \bar{u}_i \bar{u}_j}{\partial x_j} = -\frac{1}{\rho}\frac{\partial \bar{p}_i}{\partial x_i} + \nu \frac{\partial^2 \bar{u}_i}{\partial x_j \partial x_j} - \frac{\partial \tau_{ij}}{\partial x_j} + g_j \beta \bar{T} \delta_{ij} \tag{7.2}$$

$$\frac{\partial \bar{T}_i}{\partial t} + \frac{\partial \bar{u}_j \bar{T}}{\partial x_j} = \frac{\nu}{Pr}\frac{\partial^2 \bar{T}}{\partial x_j \partial x_j} - \frac{\partial \theta_j}{\partial x_j} \tag{7.3}$$

$$\tau_{ij} = \overline{u_i u_j} - \bar{u}_i \bar{u}_j \tag{7.4}$$

$$\theta_j = \overline{u_j T} - \bar{u}_j \bar{T} \tag{7.5}$$

where $\bar{u}_i$ is the component of the filtered fluid velocity in the x_i direction, $\bar{p}_i$ is the fluid pressure, ρ is the fluid density, v is the fluid kinematic viscosity, g_j is the gravitational acceleration, $\bar{T}$ is the filtered temperature, β is the thermal expansion coefficient, Pr is the molecular Prandtl number and δ_{ij} is the Kronecker symbol. τ_{ij} and θ_j are the subgrid scale (SGS) stress tensor and heat flux, respectively.

Table 7.3: Physical characteristics of inertia particles used in the simulation

Particle diameter d_p (μm)	10
Density (kg/m³)	1000
Relaxation time (s)	3.12×10^{-4}
Settling velocity (m/s)	3.06×10^{-3}

The subgrid scale (SGS) stress tensor τ_{ij} is modelled using the algebraic eddy-viscosity model proposed by Smagorinsky [5]:

$$\tau_{ij} - \frac{1}{3}\delta_{ij}\tau_{kk} = -2v_{SGS}\bar{S}_{ij} \tag{7.6}$$

where v_{SGS} is the subgrid scale velocity:

$$v_{SGS} = (C_s\Delta)^2 \left[1 - \exp\left(-\frac{y^+}{A^+}\right)^3\right] |\bar{S}| \tag{7.7}$$

Here $|\bar{S}| = \left|2\bar{S}_{ij}\bar{S}_{ij}\right|^{1/2}$, where $\bar{S}_{ij} = \left(\frac{1}{2}\right)\left(\partial_j\bar{u}_i + \partial_i\bar{u}_j\right)$ is the resolved rate-of-strain tensor. C_s is the Smagorinsky constant, its value is taken equal to 0.12. The length scale Δ is equal to the grid spacing ($h = \left(h_x.h_y.h_z\right)^{1/3}$). $\left[1 - \exp\left(-\frac{y^+}{A^+}\right)^3\right]$ is the Van Driest damping function that accounts for the reduction of the subgrid length near solid walls [6]. It is based on the dimensionless distance from the wall; $y^+ = yu_\tau/v$. u_τ is the friction velocity and A is taken equal to 25.

In the same way the heat flux θ_j is modelled:

$$\theta_j = -\left(\frac{v_{SGS}}{Pr_{SGS}}\right)\frac{\partial \bar{T}}{\partial x_j} \tag{7.8}$$

The subgrid scale Prandtl number Pr_{SGS} is taken equal to 0.5.

An unstructured grid (non-conforming embedded refinement) consisting of 1,360,000 cells was used to mesh the computational domain. The first grid point near the chamber walls and manikins at which the velocity is computed is located at $y^+ \approx 1$. Two grid points are placed within the viscous sublayer, the depth of which equals 5 wall units. In the normal-to-the-wall and normal-to-manikins directions, a non-uniform grid was employed and this in order to locate more grid-points where they are most needed. The Reynolds number for the flow at hand is 1300 based on the length of the inlet opening.

Speziale *et al.* [7] stated that a reliable LES is the one that becomes a DNS when the grid resolution is as small as the Kolmogorov scales. Thus, one cannot seek a grid independent LES, as we usually do for RANS. This is because a grid independent LES is essentially DNS, and the philosophy of LES that is based on grid dependency, loses its meaning. Celik *et al.* [8] developed a method to assess the quality of LES results. It consists of estimating an index of quality which is a measure of the percentage of the resolved turbulent kinetic energy. They stated that if more than 75% of the kinetic energy is captured, the LES is considered adequate. Pope [9] has also stated that the amount of turbulent kinetic energy that is carried by the SGS scales should not exceed 25% for the LES to be well resolved.

A simple, yet practical way of estimating the SGS kinetic energy can be made based on the assumption of equilibrium at the cut-off. In this case, the dissipation rate and the SGS kinetic energy can be evaluated as the following:

$$\epsilon_r = -\tau_{ij}\frac{\partial \bar{u}_i}{\partial x_j} = (C_s\Delta)^2|\bar{S}|^3 \tag{7.9}$$

$$k_{SGS} = C_\epsilon(\Delta \times \epsilon_r)^{2/3} \tag{7.10}$$

Typically $C_\epsilon \approx 1$ [10]. For non-equilibrium turbulent flows characterised for instance by boundary layers detachment and/or zones of strong recirculation, the assumption of equilibrium at the cut-off is not valid and it is necessary to solve a transport equation for the SGS kinetic energy to get an accurate estimation of the residual kinetic energy [11]. In this work Eqs. **(7.9)** and **(7.10)** were used to give such estimation.

A time step, $\Delta t = 0.01t^*$ was used to advance calculations. t^* is the integral time scale defined as the ratio of the inlet height to the inlet velocity. The value of the time step was imposed by the numerical stability. The LES computations are initiated from a randomly generated instantaneous inlet velocity with mean velocity and turbulent kinetic energy profiles fitted to analytical formulae [12]. The time advancement was carried out until $t = 150t^*$ to attain a flow field independence of the initial conditions. At $t = 150t^*$, residuals of Eqs. **(7.1–7.3)** became smaller than the set convergence tests indicating that the computations had reached a nearly statistically steady state. From $t = 150t^*$, the calculations were continued until $t = t = 200t^*$. In this interval, the final statistical data was collected.

7.3.2 Numerical Description of Particulate Phase

7.3.2.1 Lagrangian Approach

Aerosols are released and tracked in the turbulent flow that was described in the previous section. The physical properties of these inertial particles are summarized in Tab. **(7.3)**. Equations describing particle motion is reasonably simple as a result of the high density ratio between particle and fluid densities. Only the drag and gravity forces will be retained since other forces are in this case negligible [13]. Since the dispersion of very small particles is investigated in this work, the Brownian force should be considered. Thus, the tracking of the inertial particles within the turbulent flow obeys the following system of equations:

$$dx_{p,i} = u_{p,i}dt, \, du_{p,i} = \frac{u_{p,i} - u_{s,i}}{\tau_p}dt + n_i(t)dt - g_idt$$

$$\tau_p = C_n\frac{\rho_p}{\rho_f}\frac{4d_p}{3C_D|u_s-u_p|} \tag{7.11}$$

$$C_D = \min\left[\frac{24}{Re_p}\left(1 + 0.15Re_p^{0.687}\right), 0.44\right] \tag{7.12}$$

Here x_p and u_p are the particle position and velocity, u_s is the fluid velocity seen by an inertial particle along its trajectory, d_p and ρ_p are the diameter and the material density of inertial particles, τ_p is the particle response time, g is the gravity force per unit of mass, C_D is the drag coefficient and Re_p is the particle Reynolds number, $Re_p = dp|u_s - u_p/v|$ with v is the kinematic fluid viscosity. C_n is Cunningham slip correction factor. It is included herein to correct the drag coefficient in order to take into account the free-slip boundary conditions that occur at the surface of the particles:

$$C_c = 1 + \frac{2\lambda}{d_p}\left(1.257 + 0.4e^{\frac{-1.1d_p}{2\lambda}}\right) \tag{7.13}$$

Here λ is the molecular mean free path. $n_i(t)$ is the Brownian force per unit of mass.

$$n_i(t) = G_i\sqrt{\frac{\pi S_o}{\Delta t}} \tag{7.14}$$

Here Δt is the time step, G_i is zero-mean, unit-variance independent Gaussian random number, and S_o is spectral intensity which is computed using:

$$S_o = \frac{216\nu k_B T}{\pi^2 \rho d_p^5 \left(\frac{\rho_p}{\rho}\right)^2 C_c}$$

(7.15)

Where k_B is the Boltzmann constant, $k_B = 138 \times 10^{-25} J\, K^{-1}$

The number of particles injected was set to 500 per time step for a period of time lasting 0.1 s which corresponds to a number of iteration equal to 500 iterations. Both particle and fluid seen velocities were set equal to the fluid velocity at the secondary inlet (occupant's mouth); *i.e.* 10 m/s. Calculations were performed also with 1000 particles and only minor differences in the concentration fields were noticed ($\approx$2%). Because of the high velocity at the secondary inlet, the time step of the simulation was decreased to $t = 0.001t^*$ to keep the Courant–Friedrichs–Lewy number around 1.

7.3.2.2 Eulerian Approach

In this approach, the particulate matter phase is considered as continuum that can be described using a set of generalized equations similar to the equations used to solve the gas phase. Thus, the concept of a particulate phase consisting of individual, distinguishable droplets is abandoned. This approach is known in the literature as the two-fluid approach. If only the instantaneous mass concentration of the particulate phase C is of interest as it is the case of this study, a transport equation by turbulent motion for this property can be formulated:

$$\frac{\partial \rho C}{\partial t} + u \frac{\partial \rho C}{\partial x_j} = \Gamma \frac{\partial^2 \rho C}{\partial^2 x_j} + S_c$$

(7.16)

where Γ is the particulate matter mass diffusivity and S_c is the rate of creation or destruction of the mass concentration per unit volume. In the context of LES, Eq. (**7.16**) should be spatially filtered giving rise to the filtered transport equation for particulate phase mass concentration:

$$\frac{\partial \bar{c}}{\partial t} + u \frac{\partial \bar{c}}{\partial x_j} = \Gamma \frac{\partial^2 \bar{c}}{\partial^2 x_j} - \frac{\partial \Phi_j}{\partial x_j} + \overline{SK}$$

(7.17)

The term $\partial \Phi_j / \partial x_i$ is considered as an additional turbulent diffusional process taking place at the subgrid scales. This additional concentration flux is approximated following:

$$\Phi_j = \overline{u_j C} - \overline{u_j} \bar{C} = -\left(\frac{\nu_{SGS}}{Sc_{SGS}}\right) \frac{\partial \bar{c}}{\partial x_j}$$

(7.18)

where Sc_{SGS} is taken equal to 1. Eq. (7.18) is the filtered transport equation for the mass concentration of the fluid phase. Some of its terms should be altered when it is used to compute the mass concentration of the particulate phase.

For the case in hand, the particulate phase consists of small water droplets with a diameter typically equal to 10 μm. This has many important consequences in terms of modelling. First, the slip velocity between the fluid and the particulate phases can be confidently assumed to be negligible allowing the use of the fluid phase velocity as the convective velocity for the particulate matter concentration. For the same reason, the particulate phase mass can be assumed diffused similarly to the fluid phase mass. It is linked to the momentum diffusion through the Schmidt number Sc; $\Gamma = \nu/Sc$ with $Sc = 1$. Second, the response time and hence the settling velocity of these small water droplets is small. Still, they can have a comparable value with the turbulence time scale and convection velocity in some regions of the computation domain; near the wall for instance. Therefore, this drift phenomenon can be accounted for by (i) adding the settling velocity to the convective term to count for the drift flux caused by gravity; $v_s \frac{\partial C}{\partial x_j}$ and (ii) by considering a deposition flux on surfaces that can be portrayed by the sink term $\overline{SK}$. The sink term Sc is computed as the mass wall flux per unit of volume. It is computed for all the cells that have at least one wall face. This sink terms account for a decrease in mass concentration due to loss of mass to the walls. This sink term corresponds to particle deposition in the Lagrangian approach.

$$\overline{SK} = -v_d \times \bar{C} \times A \tag{7.19}$$

where A is the area of the wall face linked to its corresponding cell. v_d is the deposition velocity.

7.4 EXPERIMENTAL SETUP

A high quality tempered glass/stainless chamber was built for the experimental work. The materials selected are based on smooth and low electrostatic residual charge. The airflow was induced by means of a DC fan and regulated by a power supply. The inlet duct length was sized at least 60 times of hydraulic diameter of the duct to ensure that the flow is fully developed at the chamber inlet.

Mono-disperse particles of 10 μm were generated by atomization of diluted standard polystyrene microsphere suspensions (Thermo Fisher Scientific). The suspension was pre-filled to the cup container attached to a spray gun. The spray gun was fixed at the centreline (X = 0.25 m, Y = 0.33 m, Z = 0.4 m) of the head of the source. A flow regulating valve was used to conform the emission velocity to the simulation boundary conditions. The expiratory process was mimicked by a short release of particles through a spray gun connected to a compressor. The spraying duration was controlled by a timer circuit and it can be adjusted by a LabVIEW program. In this work 0.1 s was selected.

HEPA filters were installed at the inlet and the outlet to minimize the background particle count inside the chamber and to prevent cross-contamination. Type T thermocouples were selected as it has better accuracy than the other types of thermocouples. Prior to the temperature measurement, the thermocouples were all calibrated *in-situ* by 5-points measurement.

The two manikins that are made of aluminium were heated by wrapping heating wire around them. Thirteen thermocouples were used to measure inlet, manikins' surface temperature at different pre-fixed locations inside the chamber. During the entire experimental period the inlet air temperature was maintained at approximate 24 °C which is the room temperature and. The temperature difference between the inlet air and the surface temperature of the manikins was kept equal to 20 °C. All the temperatures were monitored and controlled through the same LabVIEW program.

The emission velocity was measured by a thermal anemometer (TSI, 9555) while the particle concentration was measured by an optical particle counter (TSI, 3775). Conductive sampling tube was used to sample the particles to minimize the electrostatic loss. Since there was one counter available, only a single point measurement was made at a time. Background concentration was measured 5 min prior to the start of droplets' emission. The measured concentration was subtracted from the background.

7.5 RESULTS AND DISCUSSION

Figs. **(7.1)** and **(7.2)** show the predicted time-averaged velocity contour at the mid-plane of section Y–Z and at the plane crossing the manikin closer to the air inlet respectively. A cooled-jet can be easily identified near the floor and a strong vertical buoyancy-plume is formed above the heated manikin. One key feature of displacement ventilations is the low inlet velocity. As the cold air flows around the manikin and picks up the heat, the velocity increases rapidly. It is noticed that the low momentum cooled air (0.2 m/s) near the floor level absorbs heat from the two occupants and creates a dominant vertical thermal plume in the boundary layer around the two manikins. The airflow velocity is fairly weak in all regions except inside the buoyancy plume and the bulk velocity is less than 0.1 m/s. This low velocity will influence significantly the convective transport of aerosols.

Fig. **(7.3)** shows the predicted time-averaged temperature contour along the mid-plane of Y–Z direction. This figure shows a clearer picture on how the temperature varies from the cooled-inlet to the exit. Fig. **(7.4)** shows an instantaneous velocity at the same plane. Two similar but unequal strengths of vertical plumes can be observed. Inferring carefully from both figures, it can be seen that the buoyancy plume of the heated manikin near the inlet is warmer and stronger than the other one. It is due to the coolest air contacting that heated manikin while the air contacting the other manikin has been warmed.

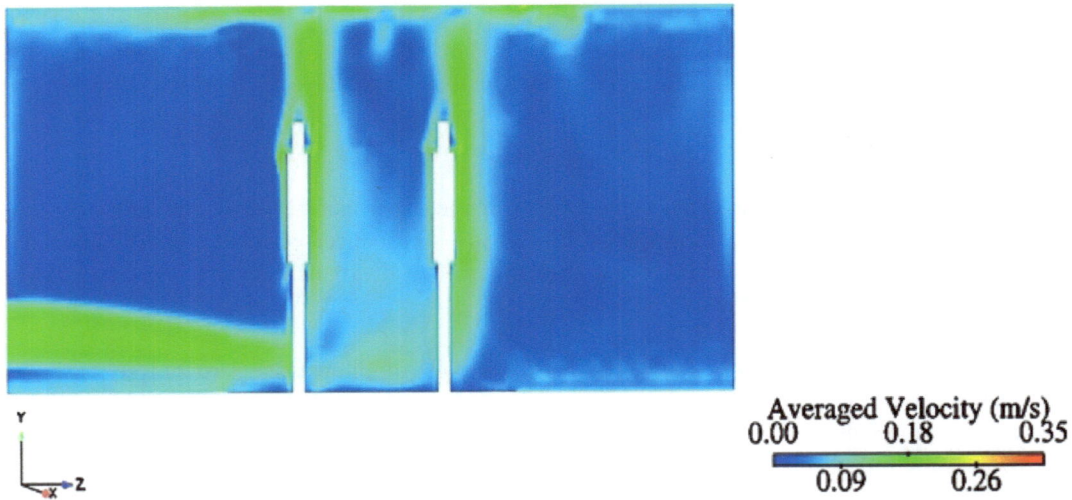

Figure 7.1: Averaged velocity contour crossing the mid-plan of section Y–Z (X = 0.25 m, Y = 0-0.5 m, Z = 0-1 m).

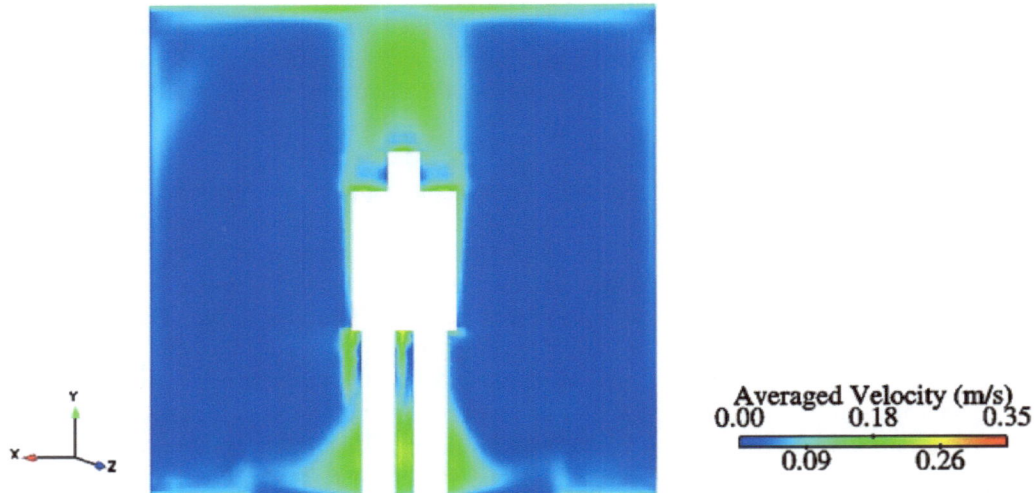

Figure 7.2: Averaged velocity contour at the section X–Y crossing the receiver (X = 0–0.5 m, Y = 0–0.5 m, Z = 0.6 m).

It is not straightforward to compare the experimental data with those modeling results. The particle counter output is number (or mass) concentration expressed in particle number (mass) per cm³. The native output for the Eulerian and Lagrangian models are concentration or dimensionless concentration. To compare with both Eulerian and Lagrangian simulation results, the (concentration) data obtained by the counter was transformed to a dimensionless concentration and is defined as:

$$C(\%) = \frac{C(x,t)}{C_{vol}(t=0)} \tag{7.20}$$

where $C(x, t)$ is the temporal concentration at the measuring point x and $C_{vol}(t = 0)$ is the initial volume-averaged concentration for the entire chamber. It was evaluated by calculating the particle injected during the emission period (0.1 s) and divided by the volume of the chamber.

Figs. **(7.5–7.7)** shows the experimentally measured steady state airflow pattern and temperature distribution prior to the emission of aerosols. The results were compared with those predicted by the LES. All the results were measured at mid-plane along Y–Z. The velocity profile along the centreline of the inlet duct (*i.e.* Y = 0.075 m) is shown in Fig. **(7.5)**. Excellent agreement between the modeling and experiment is observed. Due to the very low velocity and the sensitivity of the velocity probe used, there was no measurement data for Z greater than 0.6 m. It is interesting to note that a small recirculation zone exists at Z = 0.7–0.75 m.

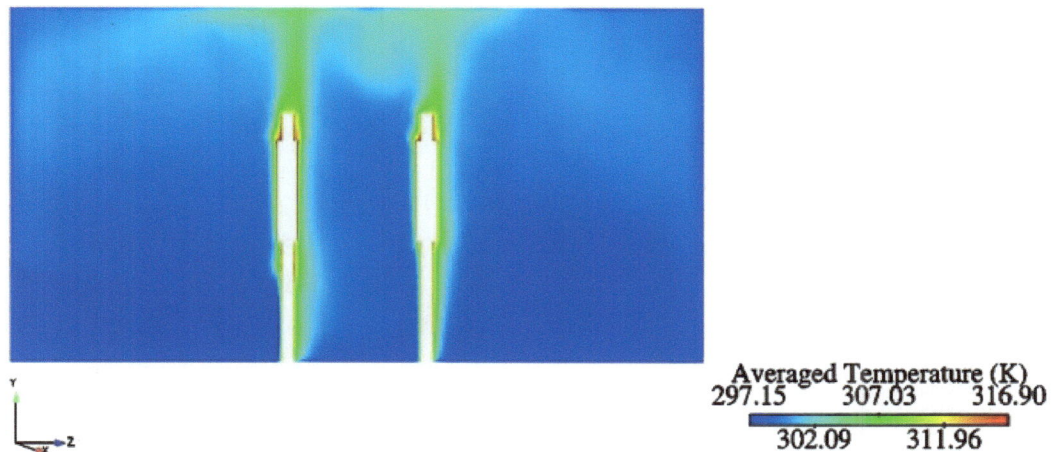

Figure 7.3: Averaged temperature contour crossing the mid-plan of section Y–Z (X = 0.25 m, Y = 0–0.5 m, Z = 0–1 m).

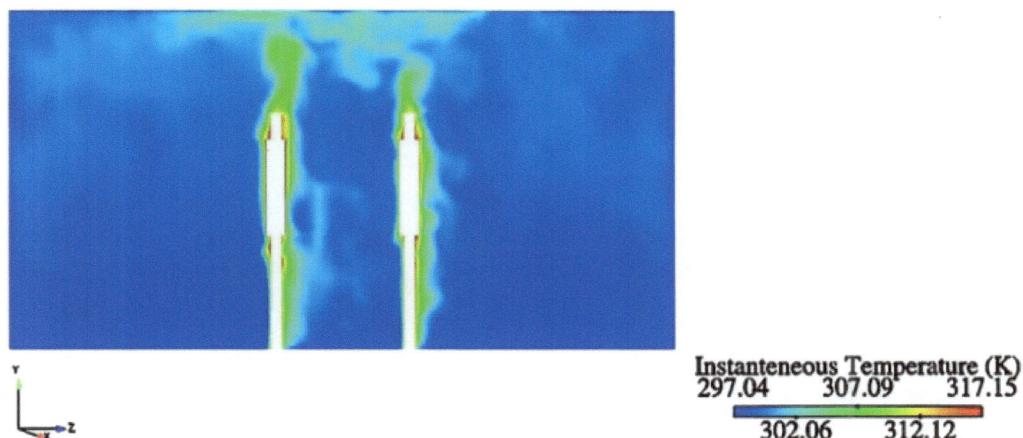

Figure 7.4: Instantaneous temperature contour crossing the mid-plan of section Y–Z (X = 0.25 m, Y = 0–0.5 m, Z = 0–1 m).

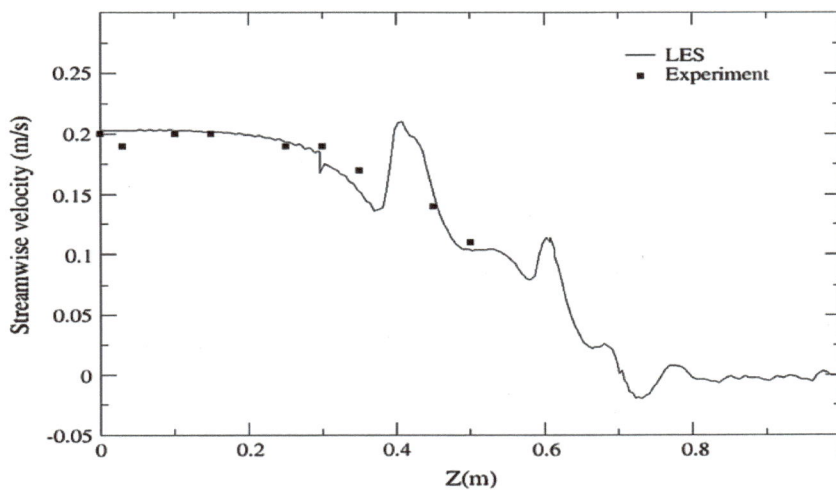

Figure 7.5: Averaged velocity distribution along the centerline of the inlet duct (X = 0.25 m, Y = 0.075 m, Z = 0–1 m).

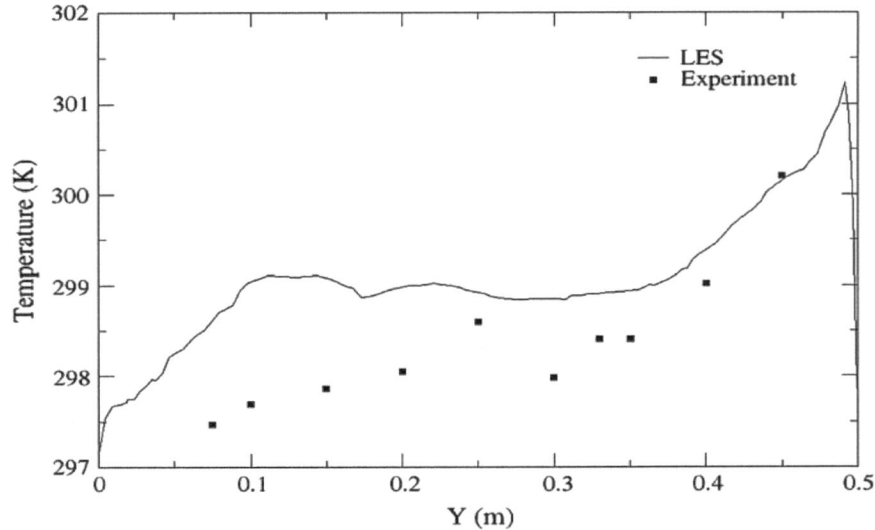

Figure 7.6: Averaged vertical temperature along the center of the chamber (X = 0.25 m, Y = 0–0.5 m, Z = 0.5 m).

Fig. **(7.6)** shows the temperature measurement along a central vertical line (i.e. Y from 0 to 0.5 m). The computational predictions consistently overestimate the measured temperature for most of the locations. It is interesting to notice that the discrepancy decreases with the Y-direction. The simulated results were fluctuated temperature while the experimental results were averaged value. Besides, the discrepancy may also be attributed to the uncertainty of the thermocouples.

Fig. **(7.7)** depicts the temperature profile along the Z-direction at the vertical height of the spray gun. Since the separation of the two manikins was less than 0.2 m, seven thermocouples were placed. It can be seen that the measurements and modelling results agree well as it shows the increasing trend when it approaches the heated manikin.

Concentration evolutions were measured in three streamwise locations for two different transverse points. Figs. **(7.8–7.11)** show the experimental results for four locations and the results are compared with the modeling predictions. The x-axis is the elapsed time after the aerosol emission and the y-axis is the dimensionless concentration. It should be noted that unlike Reynolds averaging methods, LES gives instantaneous concentrations of particles that are convected by a turbulent flow. All the reported results were measured downstream of the emission point. Hence it can be anticipated that the concentration will decay with elapsed time. Regardless of the modeling approaches or experimental results, the first three figures show clear decay profiles. For Fig. **(7.11)** where the location is at the exist, the result is more complex as both Lagrangian and experimental results depict non-decaying profiles. It can be seen that the variation of the experimental measurement at point F is very small which is ranging from 0.01% to 0.07%. Further investigation is required to explain these observations. It is also interested to compare the ''decay'' rate for different locations as the decay magnitude should be correlated to the ''local'' air velocity. This issue will be addressed in future work. The experimental results were best fitted by an exponential function using the least square method and the corresponding decay rates are shown in Figs. **(7.8)** and **(7.9)**. Inferring from these results, the decay rates are comparable. By taking into account the uncertainly of the particle counter used which is 12%, the agreement between the experimental results and the computational modeling predictions is acceptable. In this work observable discrepancy between the modeling results by Lagrangian and Eulerian approaches is seen.

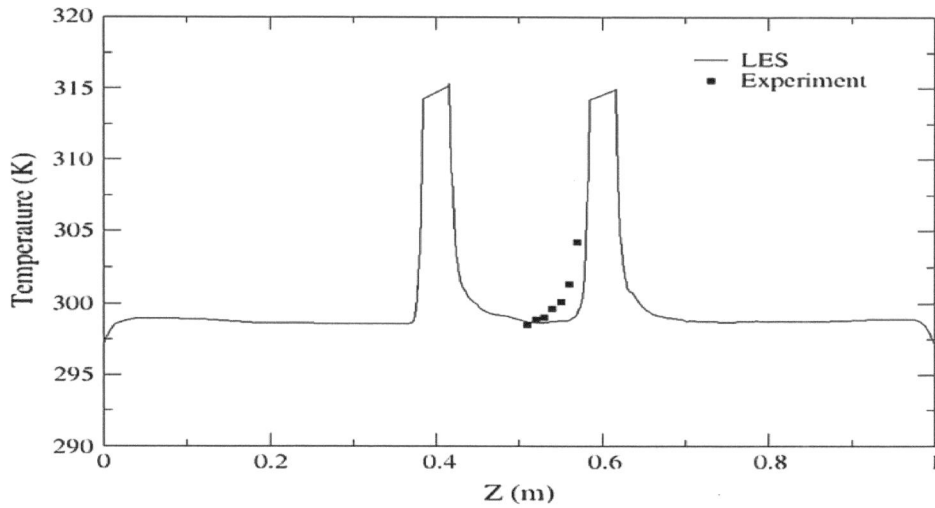

Figure 7.7: Averaged temperature distribution along the Z-direction at the vertical height of the spray gun (X = 0.25 m, Y = 0.33 m, Z = 0–1 m).

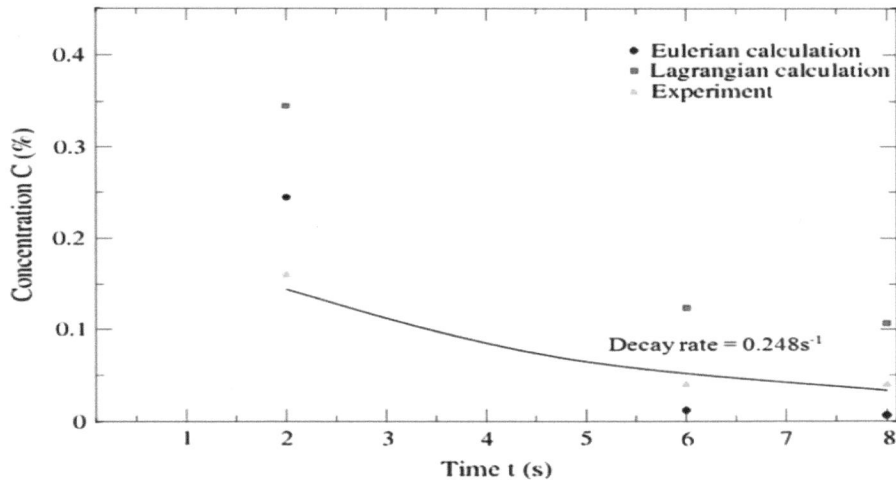

Figure 7.8: Dimensionless concentration at location C at different elapsed time after droplet emission

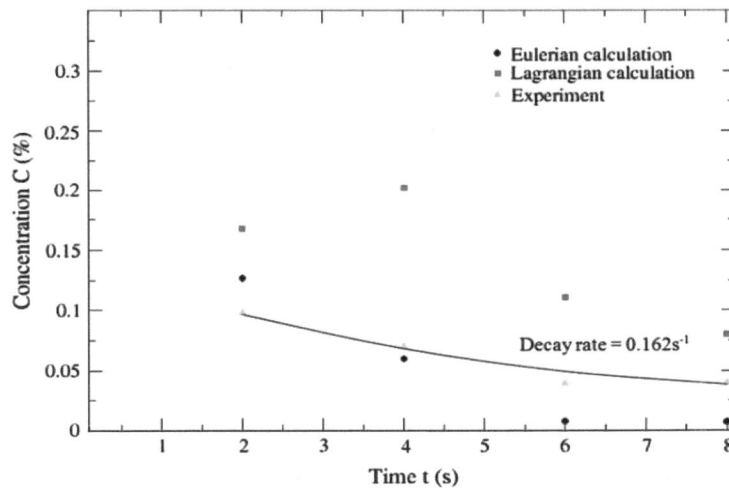

Figure 7.9: Dimensionless concentration at location D at different elapsed time after droplet emission.

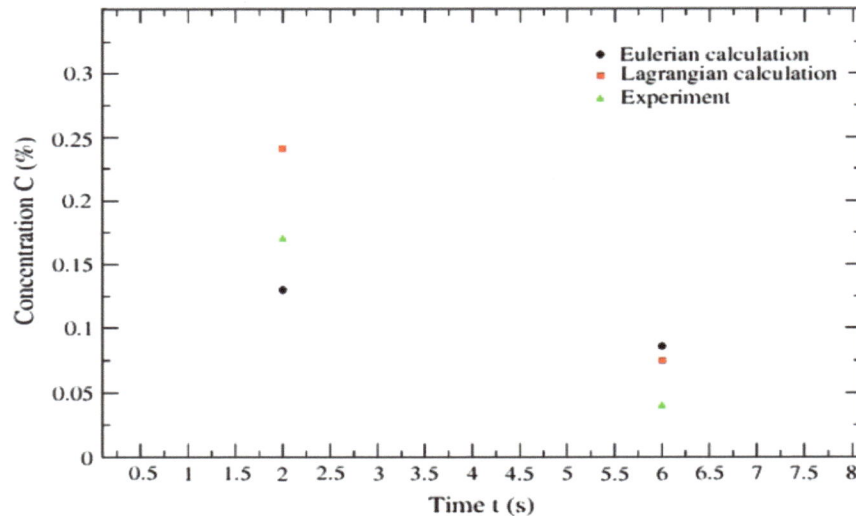

Figure 7.10: Dimensionless concentration at location E at different elapsed time after droplet emission.

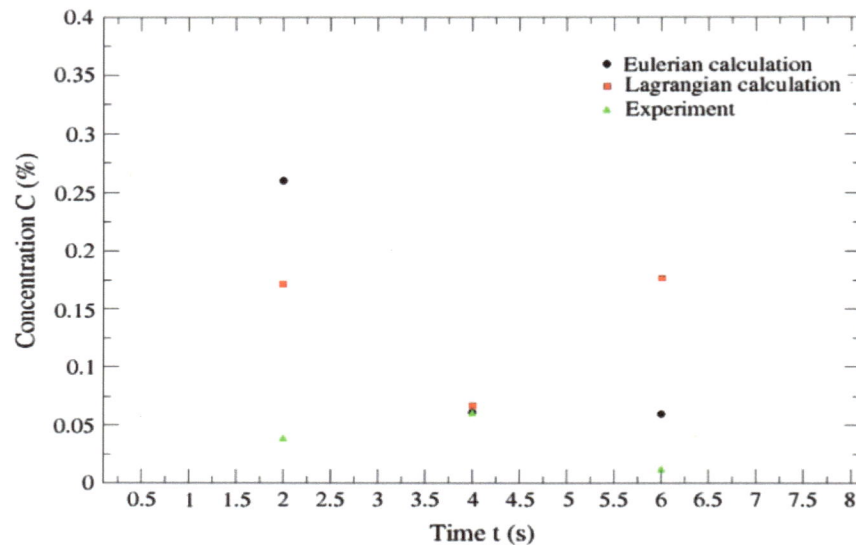

Figure 7.11: Dimensionless concentration at location F at different elapsed time after droplet emission.

REFERENCES

[1]   Zhang Z, Chen Q. Experimental measurements and numerical simulations of particle transport and distribution in ventilated rooms. Atmos Environ 2006; 40; 3396–408.

[2]   Zhang W, Chen Q. Large eddy simulation of indoor airflow with a filtered dynamic subgrid scale model. International J Heat Mass Trans 2000; 43; 3219–31.

[3]   Speziale CG. Turbulence modeling for Time-Dependant RANS and VLES: A Rev AIAA J 1998; 36(2)

[4]   Zhang Z, Chen Q. Comparison of the Eulerian and Lagrangian methods for predicting particle transport in enclosed spaces. Atmos Environ 2007; 41; 5236–48.

[5]   Clift R, Grace JR, Weber ME. Bubbles, drops and particles. London Academic Press 1978; 380pp.

[6]   Van Driest ER. On the turbulent flow near a wall. J. Aero. Sci. 1956; 23; 1007.

[7]   Crowe CT, Sommerfeld M, Tsuji Y. Multiphase flows with droplets and particles. CRC Press Boca Raton FL. 1998.

[8]   Celik IB, Cehreli ZN, Yavuz I. Index of quality for large-eddy simulations. Proceedings of ASME FEDSM2003-45448. 4th ASME JSME joint Fluids Engineering Conference. Honolulu, Hawaii 2003.

[9]   Pope S.B. Ten questions concerning the large-eddy simulation of turbulent flows. New J Phys 2004; 6; 35.

[10]   Gicquel LYM, Givi P, Jaberi FA, Pope SB. Velocity filtered density function for large eddy simulation of turbulent flows. Phys Fluids 2002; A(14)3; 1196-1213.

[11]   Moin P, Squires K, Cabot WH, Lee S. A dynamic subgrid scale model for compressible turbulence and scalar transport. Phys Fluids 1991; A3; 2746.

[12]   Laurence D. Length-scales and correlations for RANS-LES coupling Internal Report D3.3-15 for DESIDER project. University of Manchester 2005.

[13]   Maxey MR, Riley JJ. Equation of motion for a small rigid sphere in a nonuniform flow. Phys Fluids 1983; 26(4); 883-9.

Epilogue

Stochastic Large eddy simulation emerges as a promising technique for dispersed turbulent two-phase flows. A stochastic Lagrangian model based on a Langevin-type stochastic diffusion process was described in this e-Book. The primary objective of such modeling is to account for the dispersion and deposition of inertial particles by subfilter or subgrid motion that is discarded by the filtering operation in LES. Particular attention was given to the testing of the model under standard and complete formulations in shear turbulent flows taking into account inertia effects that are caused by density difference between carrier and dispersed phases and the cross-trajectory effects due to gravity.

The stochastic modeling in particle-laden LES is motivated by the inability of RANS and LES using only the filtered velocity field to properly predict transport of inertial particles with Stokes numbers smaller than the smallest resolved turbulence scales. This modeling may be crucial also for LES characterized by an excessive filtering-out of kinetic energy due to the lack of spatial resolution in regions with high shear as such near the wall and zones of recirculation and boundary layer detachment.

This e-Book highlights the progress that has been made to date to improve the predicting capabilities of the large eddy simulation technique for two-phase turbulent flows. It represents only the start of the development and validation process that should be pursued in order to confidently apply this model to an increasing number of applications of industrial and biomedical nature. I hope this e-Book conveys the potential of stochastic large eddy simulation for predicting dispersed turbulent flows and will stimulate the use of the stochastic LES to many other challenging applications.

I have made an attempt to provide sufficient information to understand and to define the approach's potential and practicality. An attempt is also made to evolve general CFD guidelines which may be useful for solving practical engineering problems using stochastic LES. Adequate attention to the key issues mentioned in this e-Book and creative use of the stochastic LES model will make significant contribution to enhancing our understanding of the subject. New advances may be assimilated using the framework discussed in this e-Book.

Appendix A

ASYMPTOTIC CASES OF THE SDE SYSTEM (4.15)

The SDE system (4.15) has a physical meaning only in the case where Δt is smaller than both the particle response time τ_p and characteristic time scale of the turbulence $T^{SGS,*}$. When this condition is not satisfied, it is possible to show that, in the continuous sense (time and all coefficients are continuous functions which can go to zero), the system converges towards several limit systems.

Here we consider:

$$\Pi(i) = -\frac{1}{\rho_f}\frac{\partial \bar{p}}{\partial x_i} + \frac{1}{Re}\frac{\partial^2 \bar{u}_{f,i}}{\partial x_j \partial x_j} + \left(< u_{p,j} > -\bar{u}_{f,j}\right)\frac{\partial \bar{u}_{f,i}}{\partial x_j}$$

Asymptotic Case 1:

$$\tau_p \to 0 : u_p(t) \to u_s(t)$$

In the discrete simulation: $\tau_p << \Delta t << T^*_{SGS}$

The limit of SDE system (4.15) is:

$$dx_p = u_s.dt$$

$$du_{s,i} = \Pi(i)dt - \left(\frac{u_{s,i}-\bar{u}_{f,i}}{T^*_{SGS}}\right)dt + \sqrt{C^d < \epsilon_r >}\, dW_i \qquad (A.1)$$

Numerical simulation: $u_p(i\Delta t) = u_s(i\Delta t)$

Asymptotic Case 2:

$$T^*_{SGS} \to 0$$

The velocity of the fluid seen has no longer a random part due to the SGS motion:

$$dx_p = u_p.d$$

$$du_p = \frac{\bar{u}_f(t,x_p)-u_p}{\tau_p}.dt + g_i \qquad (A.2)$$

Numerical simulation: $u_s(i\Delta t) = \bar{u}_f(i\Delta t, x_p(i\Delta t))$

Asymptotic Case 3:

$$T^*_{SGS} \to 0 \text{ and } C^d.\epsilon_r >> 1 \text{ such that: } C^*_0.\epsilon_r.(T^*_{SGS})^2 \to Cst$$

u_s is a fast variable and it is eliminated

In the discrete simulation $T^*_{SGS} << \Delta t << \tau_p$ and the limit of SDE system (4.15) is :

$$dx_p = u_p.dt$$

$$du_p = -\frac{u_p-\bar{u}_f}{\tau_p}.dt + \frac{\sqrt{C^d.<\epsilon>.T^{*2}_{SGS}}}{\tau_p}.dW \qquad (A.3)$$

Loosely speaking: $u_s = \bar{u}_s + \sqrt{C^d.\epsilon_r.T_{SGS}^{*2}}\,\frac{dW}{dt}$

Numerical simulation with e = N(0,1) :

$$u_s(i\Delta t) = \bar{u}_s\left(i\Delta t, x_p(i\Delta t)\right) + \sqrt{C^d.\epsilon_r.T_{SGS}^{*2}}.e$$

e = N(0,1) is a vector composed of independent normal Gaussian random variables.

Asymptotic Case 4:

$$\tau_p, T_{SGS}^* \rightarrow 0 \text{ and } C^d.\epsilon\_r >> 1 \text{ such that: } C^d.\epsilon_r.T_{SGS}^* \rightarrow Cst$$

u_s is a fast variable and it is eliminated and discrete particles behave like fluid particles

In the discrete simulation $T_{SGS}^*, \tau_p << \Delta t$ and the limit of SDE system (4.15) is:

$$dx_p = \bar{u}_s.dt + \sqrt{C^d.<\epsilon_r>.T_{SGS}^{*2}}.dW \qquad\qquad (A.4)$$

Numerical simulation with e = N(0,1):

$$u_p(i\Delta t) = \bar{u}_s(i\Delta t)$$

$$u_s(i\Delta t) = \bar{u}_s(i\Delta t, x_p(i\Delta t)) + \sqrt{C^d.\epsilon_r.T_{SGS}^{*2}}.e$$

In this case we retrieve a pure diffusive behavior that is the equation of the Brownian motion.

Appendix B

Analytical Solution of the SDE System (4.15)

The analytical solutions of the SDE system (4.15) for particle position x_p, velocity u_p and seen fluid velocity u_s are given below:

$$u_{s,i}(t) = u_{s,i}(0)exp\left(-\Delta t/T^*_{SGS,i}\right) + [\bar{u}_{f,i} + T^*_{SGS,i}\Pi_i]\{1 - exp\left(-\Delta t/T^*_{SGS,i}\right)\} + \gamma_i(t), \tag{B.1}$$

$$u_{p,i}(t) = u_{p,i}(0)exp\left(-\Delta t/\tau_p\right) + u_{s,i}(0)(\frac{1}{1 - \tau_p/T^*_{SGS,i}})\{exp(-\Delta t/T^*_{SGS,i}) - exp(-\Delta t/\tau\_p)\}$$

$$+[\bar{u}_{f,i} + T^*_{SGS,i}\Pi_i]\{1 - exp\left(-\Delta t/\tau_p\right) - (\frac{1}{1 - \tau_p/T^*_{SGS,i}})(exp(-\Delta t/T^*_{SGS,i}) - exp(-\Delta t/\tau_p))\}$$

$$+\tau_i g_i(1 - exp\left(-\Delta t/\tau_p\right)) + \Gamma_i(t), \tag{B.2}$$

$$x_{p,i}(t) = x_{p,i}(0) + u_{p,i}(0).\Delta t.[1 - exp\left(-\Delta t/\tau_p\right)]$$

$$+u_{s,i}(0)(\frac{1}{1 - \tau_p/T^*_{SGS,i}}) \times \{T^*_{SGS,i}[1 - exp(-\Delta t/T^*_{SGS,i})] + \tau_p[1 - exp(-\Delta t/\tau_p)]\}$$

$$+[\bar{u}_{f,i} + T^*_{SGS,i}\Pi_i]\{\Delta t + \tau_p(exp\left(-\Delta t/\tau_p\right) - 1)$$

$$-(\frac{1}{1 - \tau_p/T^*_{SGS,i}}) \times \{T^*_{SGS,i}[1 - exp(-\Delta t/T^*_{SGS,i})] + \tau_p[1 - exp(-\Delta t/\tau_p)]\}$$

$$+\tau_p g_i(\Delta t - \tau_p(exp\left(-\Delta t/\tau_p\right)) + \Omega_i(t)). \tag{B.3}$$

Where:

$$\Pi(i) = -\frac{1}{\rho_f}\frac{\partial\bar{p}}{\partial x_i} + \frac{1}{Re}\frac{\partial^2\bar{u}_{f,i}}{\partial x_j\partial x_j} + (<u_{p,j}> - \bar{u}_{f,j})\frac{\partial\bar{u}_{f,i}}{\partial x_j},$$

$$\gamma_i(t) = \sqrt{C^d <\epsilon_r>}\, exp\left(-\Delta t/T^*_{SGS,i}\right)\int_0^t exp\left(-t'/T^*_{SGS,i}\right)dW_i,$$

$$\Gamma_i(t) = \frac{1}{\tau_p}exp\left(-\Delta t/\tau_p\right)\int_0^t exp\left(-t'/\tau_p\right)\gamma_i(t')dt',$$

$$\Omega_i(t) = \int_0^t \Gamma_i(t')dt'.$$

Glossary

ROMAN LETTERS

$< u_i u_j >$	Reynolds stress tensor
$< u_r >$	Mean slip velocity between fluid and inertial particles
$\bar{u}$	Instantaneous filtered velocity field
$\bar{p}$	Filtered pressure field
$\overline{S_{ij}}$	Resolved rate of strain tensor
$\widetilde{kr}$	Modified residual kinetic energy
B_{ij}	Diffusion matrix
C_0^*	Diffusion coefficient
C_0	Kolmogorov constant
C_B	Scale-similar constant
C_c	Particle concentration at the pipe centre
C_D	Drag coefficient
C_n	Cunningham slip correction factor
C_s	Smagorinsky constant
C_w	WALE model constant
d_p^μ	Mean particle diameter
d_p	Particle diameter
De	Dean number
F_s	Saffman lift force
g_{ij}	Gradient velocity tensor
I	Interception parameter
k_r	Residual kinetic energy
k_{SGS}	Sub-grid scale kinetic energy
k_T	Total turbulent kinetic energy
l_d	Dissipative length scale
L_f	Fluid integral length scale
l_f	Grid filter width
l_g	Test filter width
m_p	Particle mass
N_p	Number of particles
R_0	Curvature ratio
R_b	Radius of curvature of the bend
R_{pt}	Ratio of particle size to the turbulence length scale
Re_τ	Friction Reynolds number
$Re_{p,i}$	Particle Reynolds number based on turbulence intensity
Re_p	Particle Reynolds number based on mean slip velocity
T^*	Lagrangian time scale with inertia and CT effects included
t^*	Integral time scale
T_{SGS}^*	Lagrangian sub-grid time scale with inertia and CT effects included
$T_{E,SGS}$	Eulerian sub-grid time scale
t_k	Kolmogorov time scale
T_E	Eulerian time scale

$T_{L,SGS}$	Lagrangian sub-grid time scale
$T_{L,p}$	Particle Lagrangian time scale
T_L	Lagrangian time scale
u_i'	Fluid fluctuating turbulent velocity
U_0	Mean axial velocity
u_τ	Friction or shear velocity
u_b	Bulk velocity
u_c	Centerline velocity
u_d	Settling velocity
u_p	Particle velocity
u_{SGS}	SGS or small scale velocity field
u_s	Velocity of the fluid seen
v_p'	Particle rms fluctuating velocity
V_L	Lift velocity
v_p	Particle radial velocity
W_i	Weiner process
$w_{p,c}$	Particle streamwise velocity at the pipe center
w_p	Particle streamwise velocity
x_p	Particle position
Y^+	Distance from the wall in viscous wall units
G_Δ	Filter kernel
M_i, N_i	Commutation errors
u_i	ith Velocity component
x_i	ith Cartesian coordinate
A	Drift vector
C	Particle concentration
D	Pipe diameter
Dr	Drift parameter
g	Gravity force
h	Grid spacing
I	Inner radius of the curved bend
k	LES-resolved turbulent kinetic energy
L	Integral length scale
m	Turbulence structure parameter
O	Outer radius of the curved bend
P	Mean pressure field
p	Pressure field
q	Body force per unit of mass
R	Pipe radius
Re	Flow Reynolds number
St	Stokes number
T	Integral time scale:
t	Time variable
U	Mean velocity field
y	Distance from the wall

GREEK LETTERS

Δt	Time step
α_p	Volume fraction
β	Ratio between Lagrangian and Eulerian time scales
Δ	Filter width
δ_{ij}	Kornecker Delta
ε_f	Fluid particle diffusivity
ε_p	Particle diffusivity coefficient
ε_r	Dissipation rate of the residual kinetic energy
η_k	Kolmogorov length scale
η_p	Deposition efficiency
μ	Fluid dynamic viscosity
ν	Fluid kinematic viscosity
ν_{SGS}	Sub-grid scale eddy viscosity
ν_t	Turbulent eddy viscosity
Ω	Cell volume
Φ	Ratio of eddy viscosity induced by the dispersed phase to shear-induced viscosity
ρ_f	Fluid density
ρ_p	Particle density
τ_{ij}	Sub-grid stress tensor
τ_p	Particle response time
τ_w	Wall shear stress

ACRONYMS

CFD	Computational Fluid Dynamics
CT	Cross Trajectory effect
DNS	Direct Numerical Simulation
FDF	Filtered Density Function
FV	Finite Volume
LES	Large Eddy Simulation
PDF	Probability Density Function
RANS	Reynolds-Averaged Navier-Stokes
RST	Reynolds Stress Transport
SGS	Sub-grid scale
SM	Stochastic Modeling
SOCF	Second Order Commuting Filter
WALE	Wall-Adapted Local Eddy-viscosity

Author Index

A

Akselvoll K. 15
Amiri A. E. 5
Arnason G. 4, 5, 23, 24, 36, 37, 49, 50, 57, 67, 68
Anderson R. 12, 13, 15
Apte S. V. 5
Armenio V. 10
Archambeau F. 68

B

Balaras E. 5, 15
Barbier G. 16
Bardina J. 13, 14, 15
Baytekin H.T. 94
Bellan J. 5
Benhamadouche S. 16
Berger S. A. 94
Berlemont A. 25
Bini M. 5, 15
Boersma B.J. 94
Boivin M. 5, 25, 30, 32
Boris J.P. 14
Bracco F.V. 25, 32, 33
Breuer M. 69, 72, 79, 80, 82, 84, 86, 88, 89, 90
Brun C. 16

C

Cabot W.H. 15, 107
Calabrese R.V. 34, 67
Carlier J. Ph. 28, 32
Celik I.B. 42, 68, 69, 86, 94, 98, 107
Chen Q. 96, 107
Cheng Y.S. 94
Chibarro S. 32
Chung J.N. 25
Clift R. 24, 107
Coleman H.W. 68
Collins L.R. 20, 25
Comte P. 15
Crowe C.T. 19, 20, 34, 68, 107
Csanady G.T. 28, 32, 78
Curtis J.S. 25

D

Deardorff J.W. 15
Den Toonder J.M.J. 43, 68
Desjonqueres P. 25
Deville M. 15

Subject Index

www.ingramcontent.com/pod-product-compliance
Lightning Source LLC
Chambersburg PA
CBHW041716210326
41598CB00007B/679